Common Indian
Wild Flowers

Common Indian
Wild Flowers

Isaac Kehimkar

Sponsored by the Tata Social Welfare Trust

BOMBAY NATURAL HISTORY SOCIETY
OXFORD UNIVERSITY PRESS
MUMBAI DELHI CALCUTTA CHENNAI

Oxford University Press, Walton Street, Oxford OX2 6DP

Oxford, New York,
Athens, Auckland, Bangkok,
Cape Town, Chennai, Dar-es-Salaam,
Delhi, Florence, Hong Kong, Istanbul, Karachi,
Kolkatta, Kuala Lumpur, Madrid, Melbourne,
Mexico City, Mumbai, Nairobi, Paris,
Singapore,Taipei, Tokyo, Toronto,
and associated companies in
Berlin, Ibadan

© Bombay Natural History Society
First Edition, 2000
Reprint, 2009

ISBN 019565696-2

Front and Back Covers: Graham's Groundsel *Senecio grahami*
End Papers: Reticulated bladderwort *Utricularia reticulata,* Bead Grass *Eriocaulon* sp.
Cover Design and Text Editor: Gayatri Ugra
Text and Cover layout: J.P.K. Menon

PRINTED BY BRO. LEO AT ST. FRANCIS INDUSTRIAL TRAINING INSTITUTE, MOUNT
POINSUR, BORIVLI (W), MUMBAI 400 103, PUBLISHED BY THE BOMBAY NATURAL
HISTORY SOCIETY AND CO-PUBLISHED BY MANZAR KHAN, OXFORD UNIVERSITY
PRESS, OXFORD HOUSE, APOLLO BUNDER, MUMBAI 400 039.

CONTENTS

FOREWORD

A book on wildflowers has been a long felt need. This book describes 240 of the common wildflowers seen in the countryside and the plains and forests of India, particularly the Peninsula. The author, Isaac Kehimkar, is a true naturalist, having learned his natural history in the field over a lifetime. He has multiple interests, being well versed in plants, insects, particularly butterflies, reptiles, birds, etc. He is an excellent photographer as the colour plates of the book indicate.

The book gives brief descriptions complementing the colour illustrations. The identification characters are indicated. Common names, both in English and in Indian languages where available, are given as also the distribution and status.

A very useful fieldguide to the flowers likely to be seen in the countryside.

J.C. Daniel

PREFACE

For amateurs, knowing about Indian wild flowers has been difficult in the absence of a popular guide. That is why, perhaps, the study of Indian wild flowers was not pursued by amateurs as a hobby. Here is an attempt to popularise India's wild flora, since there is at present no concise, non-specialist guide on Indian wild flowers. The book is aimed at fulfilling this long felt need. It is hoped that naturalists, botanists as well as nature photographers will find this book useful. Fortunately, a lot of literature has been published on the flora of the Indian subcontinent. Lately, some fine illustrated books on Himalayan flowers have also been published. However, so far there was no illustrated book on the flowers that one comes across on the wayside, in one's backyard, roadside and forest.

Most of the flowers shown here are native to the Indian subcontinent. However, several foreign exotics that have successfully naturalised here are also included. In fact, some exotics are more commonly seen than native plants. Typical Himalayan flowers are not included in this book. These gorgeous blooms had to be left out only after overcoming a great temptation, since there are already some very good books exclusively on Himalayan flowers. Some rare and endangered species have also been included to draw attention towards the conservation and protection of the rich biodiversity of India.

The first obstacle was to give a common English name to some of the flowers which did not have one. Besides consulting botanists and books, plant characteristics and habits were considered to coin new names. And this being the first attempt, corrections and suggestions are welcome.

The flowers have been photographed in their natural surroundings, observing the ethics of nature photography. None of the flowers were plucked or arranged to get a perfect picture. The photographs are not to scale; actual measurements are given in the text. Keeping in mind the beginner, the text has been kept simple and non-technical. Botanical jargon has been deliberately kept to the minimum. However, describing the features exactly in appropriate simple words was a problem. Therefore, only when unavoidable such words are used in the text and explained in the glossary. Herbs, shrubs and climbers have been covered, but large trees are not within the scope of this book. However, a few small trees, e.g. the painted thorn bush or false guava have been included. Species here are representative of most of the groups of flowering plants, except grasses and sedges. With the help of the illustrated species, other related species could possibly be identified. For serious amateurs, a list of references has been provided for further reading.

This book has been written for those who enjoy the beauty of flowers in the wild and are curious to know more about them. It is an effort to bring people closer to nature and let them slowly discover the wonders of nature, admire their beauty and value nature's gift to mankind.

Isaac Kehimkar

ACKNOWLEDGEMENTS

First and foremost, I wish to express my indebtedness to Mr. J.C. Daniel, Honorary Secretary, BNHS, as it was his idea to start a new series on Indian wild flowers in *Hornbill*, the Society's popular magazine. I feel most honoured that Mr. Daniel showed confidence in me to write the series, whose popularity among *Hornbill* readers finally led to the making of this book. Mr. Daniel not only corrected the draft and suggested important changes, but also closely supervised the progress of the book. Also, I consider myself fortunate to have Mr. Daniel's foreword for the book. I am most grateful to the Tata Social Welfare Trust for having funded the publication of this book. It is not just this book, but their gracious gesture has enabled the BNHS to set up the BNHS New Titles Publications Fund on a sound foundation, which will help the Society to bring out more new books on natural history. Mr. R.M. Lala, Director, Sir Dorabji Tata Trust took a keen interest in the project as the topic itself had fascinated him. I thank him with respectful gratitude. I also thank Mr. Mukund Gorakshakar, Project Officer, Sir Dorabji Tata Trust, who has been very co-operative, while he monitored the progress of this publication.

Mrs. Dilnavaz S. Variava, Vice President, BNHS took a keen interest in the project and closely followed up with the sponsors. Mr. Sunil Zaveri, Honorary Treasurer, BNHS, successfully negotiated and from time to time followed up with the sponsors, by providing the necessary information. I thank them both for their wholehearted support.

I am thankful to the Society's Publication Sub-committee for accepting my proposal for the publication of this book.

I am indeed grateful to Dr. Asad R. Rahmani, Director, BNHS for his very encouraging and appreciative support during the making of this book. Mr. Naresh Chaturvedi, Curator, BNHS very kindly corrected the text, and moreover, from time to time, provided much-needed guidance and support.

From the very beginning Dr. Gayatri Ugra, Publications Officer, BNHS has been quite enthusiastic about this project. Right from formulating the funding proposal, meeting with the sponsors, chasing me with deadlines, finalising layouts and cover design, till the final printing, Dr. Ugra personally checked every single word and photograph that went into this book.

With the same enthusiasm, the Publications team was always too happy to work on this project. Ms. Vibhuti Dedhia laid out the colour plates and ensured that colour corrections were up to the mark. Mr. Gopi Naidu gave the initial attractive look to flower plates when published as a series. For Mrs. Ivy Johnson, proof reading the text of this book was her first assignment at the BNHS, and she gave her best to this job. Mr. Jayaprakash Menon laid out the text to give its present look. I take this opportunity to express my deep gratitude to the entire Publications team, with whom I really enjoyed working on this book.

I wish to express my sincere gratitude to Mr. M.R. Almeida for having spared his time to view and check the identity of the plant pictures I have used in the book. Mr. Almeida has been like Dronacharya, from whom I have learnt identification of plants in the field, and like Eklavya, I could never be his student officially. His critical appraisal of the manuscript was indeed very valuable and helped in making necessary changes.

All through my work in the field and while writing this book, my wife Nandini has not only graciously borne the brunt of looking after the family, but she often accompanied me on my photographic trips and assisted me in the field. Her gift of a computer made writing this book more enjoyable. I am truly delighted to have such a wonderful and supportive companion. My elder son, Sameer, proved to be more than just a keen-eyed assistant, at times he helped me to climb inaccessible rock faces or straight tree trunks by becoming a human step ladder.

I thank Dr. C.S. Lattoo, former in-charge of the Herbarium at the Indian Institute of Science, Mumbai for identifying the plants and for valuable guidance.

I thank Mr. V. M. Crishna, Director, Naoroji Godrej Centre for Plant Research, Shindewadi for granting me the permission to photograph some of the rare plants at their Centre. I wish to acknowledge Dr. P. Tetali and his dedicated team at the Naoroji Godrej Centre for Plant Research for their co-operation and assistance.

I am grateful to Dr. S.M. Almeida, Director, Blatter Herbarium, St. Xavier's College, Mumbai for identifying some of the plants.

I wish to acknowledge my former colleague and botanist at the Society, Ms. Neelam Patil, who helped me to identify plants in the field. Similarly, I am very grateful to my colleague, Ms. V. Shubhalaxmi, Education Officer at the Society's Conservation Education Centre. Being a field researcher herself, she kept me posted about the plants flowering in the field and helped me to locate certain species, which were hard to find. I would like to mention my friend, Mr. Sudhir Sapre, who has been with me on several field outings and looked after me when I visited his farm at Phaltan in search of certain typical species of the drier region.

I shall always remain indebted to my parents for allowing me to do what I enjoyed most.

Finally, I owe my deepest gratitude to the Bombay Natural History Society, where I learnt the basics of natural history and was fortunate to have the opportunity to write this book.

Isaac Kehimkar

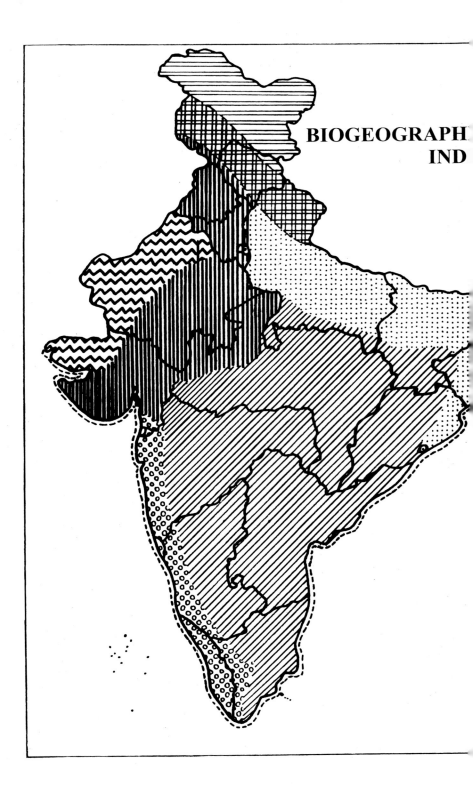

BIOGEOGRAPH
IND

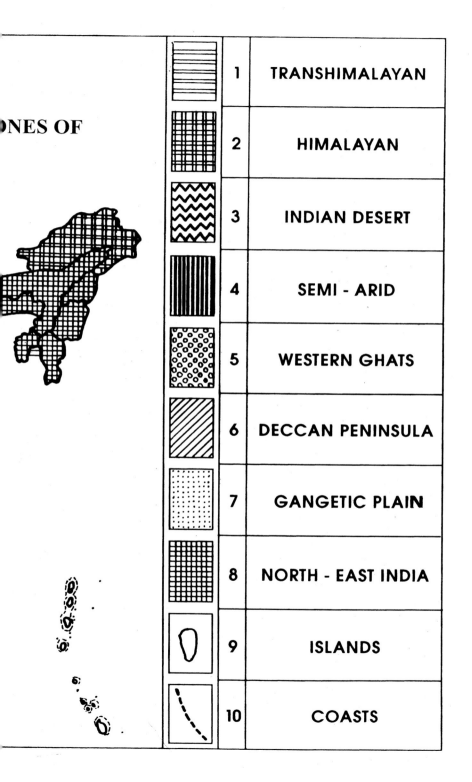

	1	TRANSHIMALAYAN
	2	HIMALAYAN
	3	INDIAN DESERT
	4	SEMI - ARID
	5	WESTERN GHATS
	6	DECCAN PENINSULA
	7	GANGETIC PLAIN
	8	NORTH - EAST INDIA
	9	ISLANDS
	10	COASTS

Abbreviations

Beng:	Bangla
Guj:	Gujarati
Hin:	Hindi
Kan:	Kannada
Kash:	Kashmiri
Mal:	Malayalam
Mar:	Marathi
Ory:	Oriya
Punj:	Punjabi
Raj:	Rajasthani
Sans:	Sanskrit
Tam:	Tamil
Tel:	Telugu

INTRODUCTION

The earliest plants on Earth were without flowers and did not produce seeds. Instead they multiplied by shedding spores. These simple plants like ferns, mosses and algae exist even today. Later, with the evolution of plants, seeds were developed, but without a protective covering, as in conifers. That is why such plants are called Gymnosperms, which means "naked seeds". These seeds are not enclosed inside an ovary, but are formed inside cones. A cone is a tight cluster of modified leaves. Usually male and female cones grow separately on the same tree. Male cones are small and produce pollen, which is released into the air to fertilize the larger female cones. On fertilization, seeds mature in the female cones.

Today, this small group has plants like pine, fir, spruce, cedar and even the palm-like cycads. Cycads are very aptly called 'living fossils' as they first appeared over 300 million years ago, and have remained unchanged ever since. Male and female cones grow on different cycad plants. However, unlike other gymnosperms, cycads can also multiply through bulbils.

Flowering plants had their beginning in the Cretaceous period of the Mesozoic era around 135 million years ago. They flourished in the Cenozoic era around 70 million years ago, which also saw the evolution of mammals and birds. In the course of evolution, flowering plants proved to be the most successful. They could grow in almost every kind of habitat on earth. Flowering plants have evolved to carry out sexual reproduction through their flowers that produce seeds. The seeds develop inside a protective structure called the ovary. Thus the flowering plants get their name Angiosperms which means "vessel seed".

Today flowering plants are the largest dominant group in the plant kingdom. India, being a vast country with wide contrasts in physical features and climate, possesses one of the richest and the most varied floras in the world. The climatic regions of India vary from scorching sun-baked deserts to the wettest regions on earth and from tropical hot deserts to cold alpine regions. Climatic factors are indeed most vital to the composition of plant life.

Basically, the climate of a region is determined by its temperature range, precipitation and the intensity of light available. These determining factors depend on the geographical parameters like the latitude, altitude, rain shadow, geology, slope of the land, direction, nature of ocean currents, day lengths and the intensity of sunlight, resulting in different habitats ranging from forests, grasslands, and wetlands, to cold and hot arid regions. All such factors have given India an array of diverse vegetative cover.

CLIMATE

Climate is one of the key factors that influence the vegetational composition of a region. Often called a land of contrasts, India has a great variety of climatic zones. During winter, the land experiences general dryness as the flow of the surface air is from the land towards the sea. Later on, the land warms up with the summer sun, causing the inflow of moisture-laden winds from the sea to the land, which results in the onset of the southwest monsoon. This is a major decisive season in the context of plant cover.

During the monsoon, along the coast and on the hills, humidity rises as high as 80 to 100 percent. Frequent rains provide ideal conditions for plant growth, both for seasonals as well as perennials. The southwest monsoon prevails mainly from June to September. Monsoon starts rather early in the Andaman Islands and Sri Lanka, while on the mainland, it begins in the far northeast and Kerala around late May or early June. Then it moves northwards, causing rainfall in the Indian subcontinent. Rainfall varies from as high as 1,092 cm at some places in the northeast and about 890 cm in some parts of the Western Ghats to as little as 10 cm of rain for an entire year in the arid northwest region. Between the winter and monsoon, there are the hot weather months March, April, and May, and soon after the rains, October. Humidity in winter and summer is much reduced. Himalaya gets summer rain, when monsoon clouds from the Bay of Bengal and the Arabian Sea cover the eastern region first, causing more rains than in the western Himalaya. Rain shadow areas shielded by mountain barriers receive scanty rain. Relatively, the western Himalaya gets more rain during the winter.

Regions like the Thar desert in the northwest experience extreme temperatures, going up to 50 °C in summer, while winter temperatures drop to freezing point. Arctic conditions prevail almost throughout the year in some of the Himalayan and Trans-Himalayan regions. On account of their proximity to the sea, coastal areas experience less variation in temperature, resulting in a more equable climate with an average temperature about 27 °C, while the Deccan peninsula and central India have extreme temperatures. The Western Ghats form a massive barrier which prevents most of the southwest monsoon from reaching the Deccan plateau, bringing this region under rain shadow. However, during winter, this region receives rain between October and mid December. At the same time in the north, the plains receive light showers and there is snowfall on the mountains.

TOPOGRAPHY

Next to the climate, the physical features of the land are an important factor in determining its plant cover. Few countries in the world have physical features as diverse as India. Altitude is an important decisive factor to influence the vegetation of the land. In the north is the Himalaya that ranges from Kashmir to Arunachal Pradesh. It has some of the tallest mountains in the world. From the Himalaya, several rivers flow into the Indogangetic plains. In the west, the Indus and its tributaries flow in a southwesterly direction towards the Arabian Sea, while the Ganga and its tributaries drain the fertile plains of Uttar Pradesh, Bihar and Bengal before emptying into the Bay of Bengal. In the northeast, fast flowing Himalayan rivers come thundering down from the foothills and through the flood plains of Assam, to join the mighty Brahmaputra, which finally meets the Ganga as it flows towards the Bay of Bengal. The basin of the Ganga, Indus, Brahmaputra and their tributaries form the alluvial plains between the Indian peninsula and the Himalaya.

The Indian peninsula, south of these plains forms a triangular tableland intersected by rivers. Along the west coast are the Western Ghats, the mountain range which runs more or less parallel to the coast facing the Arabian Sea. Between the sea and the Ghats lies a narrow, gently sloping coastal region that receives a good amount of rain and therefore is rich in floral diversity. From the Western Ghats there are prominent offshoots like the Nilgiris. Similarly there are the Anaimalai, Palni hills and other south Indian hill ranges. While the rainy western sides of the Ghats have typical evergreen vegetation, their drier eastern slopes have semi-evergreen and deciduous vegetation. Along the east coast, there arises a small range of hills known as the Eastern Ghats. In the north is the Aravalli range and then the rising uplands of Central India. Between these hill ranges, lies the central Deccan Plateau intersected by rivers like the westward flowing Narmada, and the Godavari, Krishna and their tributaries that flow towards the Bay of Bengal.

VEGETATION

Blessed with a wide variety of physical features and climate, India has probably the richest and most diverse plant life compared to countries of a similar size. The vegetation of the Indian region is broadly classified into sixteen major types here.

The **Moist Tropical** vegetation is classified into four types. **(1) Wet Evergreen** type is seen in the south as well as in the north. In both regions, the vegetation is almost alike. Evergreen forests have typically tall straight trees, the majority of which are evergreen. These trees often have a trunk buttressed at the base like a tripod to support them during storms. Trees in this type of vegetation are festooned with mosses, ferns and orchids. Canes, screwpines, tree ferns and woody climbers (lianas) are common, and wild banana is seen on the hill slopes. In the south, this type of vegetation is seen along the Western Ghats, in the Andaman and Nicobar Islands and in neighbouring Sri Lanka. In the north, the Northeast region of India, and the neighbouring Bangladesh and Myanmar have wet evergreen vegetation. **(2) Semi-Evergreen** vegetation occurs between the tropical and moist deciduous and therefore combines both these types. This vegetation is seen on the eastern slopes of the Western Ghats, the Andamans and Nicobars, and the foothills of eastern Himalaya. **(3) Moist Tropical Deciduous** type vegetation is further divided into: (a) moist teak bearing, (b) northern moist deciduous forests and (c) southern moist deciduous forests. Typical dominant trees are red silk cotton, mango, rosewood and some teak. Most larger trees have buttressed trunks and along with other smaller trees turn leafless in the dry season. The undergrowth is of bamboo, climbers and other shrubs. A good number of orchids too are seen in these forests. This type of forest vegetation occurs throughout India in suitable habitats. **(4) Littoral and Swamp** type of vegetation occurs in the coastal region along the river estuaries, creeks, inlets and islands. Such vegetation is seen in three zones: (a) Mangroves are seen in saline swamps. These salt-tolerant plants have breathing roots, an

3

adaptation for aeration. On the eastern coast, mangrove, date and nipa palms are prominent among the tidal forests of West Bengal and Orissa. Bulletwood (*Manilkara* sp.) occurs in the littoral forests of the Andaman and Nicobar Islands. (b) Vegetation of coastal sand dunes has plants like *Crotalaria verrucosa, Launaea pinnatifida,* mangrove beanstalk (*Derris trifoliata*), and goat-foot glory (*Ipomoea pes-caprae*). Shrubs like screw pine (*Pandanus* sp.) are typical. (c) Vegetation beyond the high tide mark is more or less similar to the adjoining mainland.

Dry Tropical vegetation is classified into three types: **(5) Deciduous** vegetation occurs throughout the northern region, where dominant trees and shrubs are leafless in the dry season. Typical deciduous trees are *sal* (*Shorea robusta*), axlewood (*Anogeissus latifolia*), khair (*Acacia catechu*), riverine sisso (*Dalbergia* sp.). Some conspicuous climbers, shrubs and herbs are crab-eyed creeper (*Abrus precatorius*), camel-foot climber (*Bauhinia vahlii*), spiny asparagus (*Asparagus racemosus*), touch-me-not (*Mimosa pudica*), pot cassia (*Cassia tora*) and common tephrosia (*Tephrosia purpurea*). In the southern peninsula, while some deciduous forests are dominated by teak, in other forests teak is absent. Here axlewood is common, associated with *ain* (*Terminalia* sp.), *tendu* (*Diospyros* sp.), incense tree *Boswellia serrata* and the *karaya* or ghost tree (*Sterculia urens*). Except in the Western Ghats, this type of vegetation is seen in Maharashtra, Gujarat, Madhya Pradesh, Andhra Pradesh, Karnataka, and Tamil Nadu. **(6) Thorn** forests are typical in drier regions of north, west, central and southern India. Spiny acacias with low crowns dominate the landscape, with thorny spurges (*Euphorbia* spp.) and capers (*Capparis* spp.). **(7) Dry Tropical Evergreen** type vegetation is confined to the coast of Karnataka, Tamil Nadu, Andhra Pradesh and neighbouring Sri Lanka. Typical trees are the hard-leaved evergreen trees like *Manilkara hexandra,* the *maulsari* or *bakul* (*Mimusops elengi*), *Memecylon edule* and *jamun* (*Syzygium cumini*).

Montane Subtropical vegetation is classified into three types: **(8) Broad-leaved** vegetation occurs in southern and in northern forms. In the south, this type is confined between 1,000 and 1,700 m in the south Indian hills. The northern form is seen on the lower slopes of the eastern Himalaya from 1,000 to 2,000 m. Typical trees are alder, oaks, chestnuts, cherry, magnolias, laurels and birches covered with mosses and ferns. This region probably receives the highest amount of rain. **(9) Pine** type vegetation in the western Himalaya is recognized by the dominance of *chir* pine. The pine forests occur more commonly at 1,000-1,800 m from Western Himalaya through Central Himalaya up to Sikkim. At places, pine vegetation occurs as low as 600 m and as high as 2,300 m. In moist situations, the broad-leaved oak is common. Tree rhododendron (*Rhododendron arboreum*) and wild pear (*Pyrus pashia*) are also seen. Shrubs like musk rose (*Rosa moschata*), raspberry (*Rubus* sp.), wig plant (*Rhus* sp.) occur commonly. In the northeast, *chir* is replaced by the khasi pine in the Khasi, Naga and Manipur hills at the same altitude.

On the lower hills in Western and Central Himalaya and on the plains, Indian laburnum (*Cassia fistula*), amla (*Emblica officinalis*), sal (*Shorea robusta*), flame of the forest

(*Butea monosperma*), red silk cotton tree (*Bombax ceiba*), *shisham* (*Dalbergia latifolia*), *jamun* (*Syzygium cumini*), toon (*Toona ciliata*), *haldu* (*Haldina cordifolia*) are some of the common forest trees. The upper limit of subtropical zone is at about 1,500 m, after which moss and fern covered oaks, magnolias, laurels and birches replace *sal* and silk cotton trees.

(10) Dry Evergreen type of vegetation has wild pomegranate, and small evergreen trees e.g. olive. Another common evergreen shrub is *Dodonea viscosa*, which has bright green, shining, leathery leaves. This vegetation is typical of inner dry valleys and open hillsides from Afghanistan to western Nepal, which have a long, hot and dry summer, and a cold winter.

Montane Temperate vegetation is classified into three types. **(11) The Montane Wet Temperate** type of vegetation occurs in the South Indian hills and Sri Lanka as well as in the Himalaya. In the south, **Southern Montane Wet Temperate** type is seen in small, evergreen pockets called shola forests in the Nilgiris. Typical trees of this region are Nilgiri rhododendron (*Rhododendron arboreum* ssp. *nilagiricum*), Nilgiri champa (*Michelia nilagirica*) and holly (*Ilex* sp.). Tree ferns (*Cyathea gigantea* and *C. brunoniana*) and woody climbers are common. Mosses, ferns and orchids cover the tree trunks. These unique forests occur at 1,500 m and above in sheltered patches surrounded by rolling grasslands in the Nilgiris, and the high ranges of Kerala and Sri Lanka. Rainfall ranges from 150 to 625 cm and the maximum rainfall is in October.

The **Northern Montane Wet Temperate** type of vegetation occurs from Eastern Nepal to Arunachal Pradesh between 1,800 to 3,000 m; here a minimum of 200 cm of rain is received in the monsoon, when the forest is covered with dense mist. It is relatively dry during winter from November to March. Mainly conifers, associated with oaks, form a large tree canopy. In addition, there are laurels, magnolia, rhododendrons, champa, cherry, maple, tree ferns and bamboo. Trees covered with mosses, ferns, and orchids are common. **(12) Montane Moist Temperate** type of vegetation is seen along the Himalaya at altitudes between 1,500 and 3,300 m. In the Western Himalaya, this type of vegetation is composed of broad-leaved oaks, walnut, rhododendrons and conifers. At the lower altitude, grey oak (*Quercus leucotrichophora*), *chir* pine (*Pinus roxburghii*) and alder (*Alnus nepalensis*) are typical. On the higher reaches there are trees like blue pine (*Pinus wallichiana*), western Himalayan fir (*Abies pindrow*), deodar (*Cedrus deodara*), horse chestnut (*Aesculus indica*), Himalayan cypress (*Cupressus torulosa*) and yew (*Taxus baccata*). The shrub flora is rich and interesting. Among shrubs *Clematis* (climbers), *Indigofera, Viburnum, Skimmia* and *Daphne* are more common. In the eastern Himalaya, besides the broad-leaved oaks, maple, birch, magnolia, rhododendrons, conifers like hemlocks, East Himalayan fir or red fir (*Abies densa*) are seen along with bamboo and ferns. Shrubs like raspberry, barberry, daphne and *Piptanthus nepalensis* are common. Here the rainfall is heavy, so the diversity of species is high. Conifers like *Cephalotaxus, Podocarpus*, east Himalayan spruce, larch and silver fir are found only here. **(13) Montane Dry Temperate** is predominantly coniferous, along with broad-leaved trees like oak, ash and maple. There are also chilgoza pine (*Pinus gerardiana*), deodar, juniper, high level fir (*Abies spectabilis*) and silver birch or bhojpatra (*Betula utilis*). This type of vegetation is seen in Kashmir, Lahaul, Kinnaur and Sikkim. Here in the inner ranges, precipitation does not exceed 100 cm and is received mainly as snow in winter.

(14) Subalpine vegetation is seen at altitudes from 2,900 to 3,500 m throughout the Himalayan range, from Kashmir to Arunachal Pradesh. In the Western Himalaya, five species of rhododendrons are seen along with junipers, black currant and willows. This is the region to see the gorgeous Himalayan blooms of blue poppies, primulas, potentillas and brahma kamal *(Saussurea obvallata)*, which occur in the Eastern Himalaya too. In the Eastern Himalaya, shrubs like red fir and black juniper are common. But the most spectacular are the rhododendrons. Here 82 species of rhododendrons are known to occur and most of them are shrubs. Among the notable species are *Rhododendron dalhousiae* that has white fragrant flowers, *R. thomsonii* with blood red flowers and *R. cinnabarinum* with brick-red flowers.

Alpine vegetation occurs between 3,650-5,500 m, in two forms. **(15) Moist Alpine** vegetation has low, dense, scrub evergreen forest, almost entirely rhododendron, some birch with scattered patches of fir and pine. The wet ground is covered with mosses, ferns and herbs. The plant life depends entirely on snowfall and melting snow. As the snow melts, heralding spring, clear patches get carpeted with an array of flowering herbs like anemones, primulas, *Corydalis, Fritillaria* and several more. Beyond 5,500 m, plant life is almost absent. Snow and ice is ample and permanent.

(16) Dry Alpine vegetation is dominated by shrubs like common and black juniper, and other plants like honeysuckle and potentillas. Depending upon the location and aspect, the upper alpine zone 1,000 m onwards is often arid, bare and steppe-like. Of course, similar steppe-like conditions may prevail at much lower altitudes, localised in secluded valleys that come under the rain shadow.

BIOGEOGRAPHIC AREAS

Himalaya

The Himalayan mountain range has the highest mountain peaks in the world, and profoundly affects the climate and vegetation of almost the entire Indian subcontinent. The Himalayan range has diverse flora, more or less distinct from the rest of India. Three types of vegetation are seen. Sub-tropical vegetation is found in the lower foothills and warmer valleys. Higher lies the temperate zone, which extends upwards to the tree-line, that is the upper limit of tree vegetation all along the Himalaya. From the tree-line upwards is the alpine vegetation, which occurs up to the zone of permanent snow and ice at the higher reaches. The lower region of the alpine zone has alpine meadows, known for their gorgeous variety of bright colourful flowers. Higher up, there are alpine deserts with little or no vegetation due to the peculiar geographic and climatic conditions. Between these types there is transitional overlapping.

The eastern Himalayan region comprising of Central Nepal, North Bengal, Sikkim, Bhutan up to Arunachal Pradesh, has rich and diverse flora. Its geographical position, being the nearest mountain barrier to face the moisture laden southwest monsoon wind from the Bay of Bengal, makes it the most humid part of the Himalaya. The average annual rainfall is often more than 200 cm. The climatic diversity in this region is indeed unique. At places, within a vertical distance of about 100 km, there are orchids, woody climbers (lianas), wild banana, tree ferns and screw pine (*Pandanus* sp.) in the humid tropical forest, and then the rhododendrons, laurels and conifers of the temperate forest, to the cold arctic alpine zone. Strangely, in this very region there are almost rainless, dry valleys, due to the rain shadow effect. The eastern region is tropical, while the extreme northwest of the western region has temperate vegetation.

The Western Himalaya extends from West Nepal through Uttar Pradesh (Kumaon) to Kashmir and beyond in Pakistan. The tree-line is much lower in the Western Himalaya at 3,650 m, than in the Eastern Himalaya, where it is at 4,570 m. The region comprising of northern and western Pakistan and Jammu and Kashmir, separated from the rest of the Himalaya by the Sutlej river, is referred to as the Trans-Himalaya. Zanskar, Ladakh, and Karakorum dominate this region. Vegetation includes subtropical, evergreen, and coniferous forests, and alpine steppe.

Northeast Region

The Northeast region comprises of Assam, Nagaland, Manipur, Mizoram, Meghalaya and Tripura, with southeastern Arunachal Pradesh, north Bengal (Darjeeling region) and the neighbouring northeastern and southwestern Bangladesh. The average annual rainfall often exceeds 200 cm, and supports the most rich and diverse flora and fauna in the Indian Subcontinent. It extends north towards the foothills of Himalaya from the northern fringe of the plains of Bengal. Heavy rainfall and dense, humid evergreen and semi-evergreen forests are characteristic features of this region. Forests here are storeyed and have a similar gradation from tropical towards moist temperate vegetation. However, alpine flora of the higher Himalaya is absent from the hills of this region. Interestingly, several trees and shrubs here are identical or closely related to those in the Nilgiris.

In the tropical zone, besides several tree species like the towering hollong (*Dipterocarpus macrocarpus*) and camphor (*Cinnamomum*), several evergreen shrubs and woody climbers, tree ferns, screw pine, wild banana, giant bamboo and ferns are conspicuous in the hill forest, which are home to several gorgeous orchids. The insectivorous pitcher plant (*Nepenthes khasiana*) is endemic to the Khasi hills.

Higher on the hills in the moist temperate zone, several species of oak, laurel, magnolia, rhododendron, champa, cherry, maple, tree fern and bamboo occur.

Northwest Region

Predominantly arid and semiarid tracts of the Indus Plain, this region includes Punjab, Haryana, Rajasthan west of the Aravalli range and Yamuna river, Kutch region of Gujarat, and also Sind and Baluchistan in the eastern region of Pakistan. In this region the Thar desert spreads over 446,000 km², while the rest is composed of hills and stony plateaus. Besides the Indus and its tributaries, a number of large rivers and an extensive canal system are present. Annual rainfall ranges from 25-50 cm, with the dry season lasting from 8 to 10 months. In some regions, the annual rainfall is even less than 25 cm. Due to the dry climate, this region has a scanty cover of natural vegetation. Forests in this region are usually stunted, except towards the lower altitudes of the Western Himalaya and the slopes of the Aravalli range. A large tract of land of this region is saline or sandy, and therefore plants are typically those that grow only in saline or sandy soils. These plants are hardy, drought resistant and adapted to live in harsh, arid conditions. Most of the deep-rooted perennial plants occur in open clump formation, with plenty of vacant space between them. Seasonal annual plants constitute the majority of the vegetation of this region. Annuals appear suddenly just after the first rains and complete their life cycle in a very short time. Tenacious trees like *Acacia* spp., khejari (*Prosopis cineria*), Rajasthan teak (*Tecomella undulata*), pilu (*Salvadora oleoides*) and wild date (*Phoenix sylvestris*) are typical in this region. Hardy shrubs like the lesser milkweed (*Calotropis procera*), capers (*Capparis* spp.), rock spurge (*Euphorbia caducifolia*) and ber (*Ziziphus* spp.) grow commonly in clumps with other plants. Much of this region is now under intensive agriculture, as a result the flora has lost its natural habitat.

Gangetic Plain

This region extends from the plains of western Uttar Pradesh to the plains of the northeast drained by the Ganga and its tributaries in the Hooghly delta and adjacent littoral forests of the Sunderbans. In the northwestern region of the plain, plant life is more like the desert type, but eastwards the vegetation becomes more luxuriant. In the western tract, *khejari,* babul (*Acacia* spp.) and painted thorn bush (*Dichrostachys cinerea*) are seen, while in the eastern side towards Bengal, trees like golden champa (*Michelia champaca*), and red silk cotton are common. Much of the original vegetation of this area has been greatly altered due to intensive agriculture, except in the estuarine region of the Sunderbans. However, with an abundance of wetlands and other water bodies, it is rich in aquatic or semi-aquatic vegetation. In the Sunderbans, mangrove forests attain maximum growth as they receive a higher amount of rain than India's other coastal regions.

Central India

Between the Gangetic plain and peninsular India are the plains of Central India. In this region, two great hill ranges – the Vindhya and Satpura – are situated. The Pachmarhi plateau, nestled among the Satpura hills, has very interesting plant life. Trees like teak, bahera, sal, rosewood, mahua, bonfire tree, red silk cotton, yellow silk cotton, and flame of the forest are most conspicuous in this region. Among the sal forests, the male bamboo (*Dendrocalamus strictus*) and the camel-foot climber (*Bauhinia vahlii*) are very common. The Pachmarhi hills have a rich and diverse flora of shrubs, climbers and herbs like common conehead (*Carvia callosa*),

8

silk cotton, and flame of the forest are most conspicuous in this region. Among the sal forests, the male bamboo (*Dendrocalamus strictus*) and the camel-foot climber (*Bauhinia vahlii*) are very common. The Pachmarhi hills have a rich and diverse flora of shrubs, climbers and herbs like common conehead (*Carvia callosa*), Malabar blackmouth (*Melastoma malabathricum*), Deccan clematis (*Clematis triloba*), balsams, sundew plants (*Drosera* spp.), orchids and ferns. In the drier tracts, babul (*Acacia nilotica*) is seen along with khejari in the thorn forests.

Peninsular India

South of the Indogangetic plains is the Indian peninsula, forming a triangular plateau of wide, undulating plains separated by ranges of flat-topped hills. This region is comprised of Central India, the Deccan, Western and Eastern Ghats. As the Western Ghats block the southwest monsoon, they get the full benefit of the rainfall. The plateau is intersected by several rivers flowing towards the east coast, and tapers on the east into a low range of hills called the Eastern Ghats.

Western Ghats

The Western Ghats or Sahyadri is the mountain range running parallel to the west coast from south Gujarat to Kanyakumari, covering south Gujarat, western Maharashtra, Karnataka, Kerala and Tamil Nadu. The Southwest monsoon brings in heavy rainfall to the narrow strip of the west coast and western slopes of the Ghats. Next to the northeastern region, this region has the richest diversity of flora and fauna. Luxuriant vegetation covering the western slopes is at places very dense, with tall trees, woody climbers, bamboo, wild banana, canes, ferns and orchids. Shrubs like coneheads (*Carvia* sp.) and *Ixora* sp. are typical of this region. Annuals like balsams, smithias and groundsels (*Senecio* spp.) are conspicuous as the monsoon ends. Tropical rain forests and moist-deciduous forests occur below 1,500 m, while temperate vegetation is found above 1,500 m.

Nilgiris – Palni

Among the Nilgiris, Anaimalai and Palni hills, the Nilgiris form the apex of the Western Ghats in south India. Here the peaks are gently sloping, with large grassy areas having sheltered dense evergreen forests in the gorges called *Sholas*. Annual rainfall is usually in excess of 200 cm. An interesting feature of these hills is the presence of the Nilgiri conehead (*Nilgirianthus* sp.), which flowers *en masse* every 3 to 12 years at regular intervals. The blue flowers of this shrub give the Nilgiris their name which means 'Blue Mountains'. Interestingly, the vegetation of the Nilgiris is very similar to that of the Khasi, Naga and Manipur hills.

Deccan

East of the Western Ghats in Peninsular India is an elevated hilly plateau of medium height called the Deccan region. The plateau, sheltered from the Southwest monsoon

9

Eastern Ghats

The Deccan plateau tapers off to the east into a range of low hills known as the Eastern Ghats. These isolated hills run from Orissa southwards to central Tamil Nadu, where they turn southwest to meet the Western Ghats in the Nilgiris. The steeper slopes of the Eastern Ghats have dense forests, due to the benefit of the northeast monsoon. The dominant vegetation is dry-deciduous, with patches of moist-deciduous and semi-evergreen forests. The northern and the southern sections of the Eastern Ghats are separated by the Godavari delta and other such breaks by the Mahanadi and Krishna rivers. Between the Eastern Ghats and the sea is a lowland known as the Coromandel coast. The Coromandel region has a distinct flora, along with those plants that occur on the Deccan plateau. Good mangrove forests are seen along the estuaries. Western Coromandel, that borders on the Eastern Ghats, has thorny evergreen and deciduous trees and shrubs. Further south, it is extremely hot and dry with a vegetation of the semi-arid type.

Andaman and Nicobar Islands

The Andamans are a group of 204 islands and the Nicobars of 22 islands in the Bay of Bengal. Being home to several rare and endemic flora and fauna, these islands are very important and fragile biogeographic regions. As much as ten percent of the flora of these islands is endemic, including 225 species of vascular plants. The Andaman and Nicobar Is. often receive more than 300 cm total rainfall over almost half the year. Being comparatively remote, the forest is naturally protected, though motorable roads and settlers will possibly not let the forest last long. Surrounded by the sea on all sides, the mangrove vegetation is well developed, and includes the gregarious stemless palm (*Nypa fruticans*) and betel nut palm (*Areca triandra*).

Along the coast, a tall, protective stand of the bullet wood (*Manilkara littoralis*) takes the full force of the southwest monsoon. The screw pine (*Pandanus* sp.) is common on the coastal fringes. Other prominent tall trees are gurjan (*Dipterocarpus alatus*) and padauk (*Pterocarpus dalbergioides*). Forests composed of evergreen, semi-evergreen and deciduous vegetation are rich in bamboo, cane, ferns and orchids. Vegetation on the Nicobars is very similar to that of the Andamans. The Great Nicobar is the largest island among this group. Tree ferns (*Cyathea* sp.) are common in the moist valleys of this island, which has freshwater springs, streams and rivers.

WATCHING WILD FLOWERS

Watching wild flowers can be one of the most enjoyable pursuits, especially if you enjoy sunshine and walking. You can watch flowers at leisure and unlike bird watching, you need not be out before sunrise. Most flowers bloom around sunrise, while some are nocturnal or crepuscular. Even while travelling, railway tracks and roadsides can be most productive.

Time is of the essence in finding flowers. Most morning glories (*Ipomea*), for example, bloom around dawn and by 10 a.m. they fade, while some like the day glories (*Merremia*) will not bloom before 9 a.m. and some like night glory (*Rivea*) bloom around sunset and by sunrise they fade. The tiger's paw glory will not bloom till the sun is in the west, and if you reach a lily pond after 11 a.m. for red water lilies, you will find none open. Then only the next morning will be fruitful. Similarly, flowers of the forest triumfetta (*Triumfetta rhomboidea*) will not bloom till late in the day. Such are the different time slots one has to be aware of. For Amaryllid or forest spider lilies, lose a day here or a day there, and you may have to wait for another year, as you will miss their ephemeral blooms that can be seen only during the first few weeks of the monsoon. Some plants like the forest beanstalk (*Derris scandens*) have an extended flowering period, however, within this period there are several peak periods of profuse flowering. If you miss the first peak flowering of the season, watch out for the second flowering to follow soon.

Watching flowers on the hills is very rewarding, both before and after the rains. From February to May, several perennial shrubs, climbers and trees flower in preparation to have seeds when it rains. The departing monsoon leaves a colourful trail of monsoon flowers swaying on the slopes. Walking on the hills among these flowers is a source of endless joy. But there are some like the dragon stalks (*Amorphophallus* sp.), cobra plant (*Arisaema* sp.) and several lilies which shoot out their flower from the dry forest floor as the pre-monsoon clouds begin to gather. Watching flowers in the Himalaya is indeed a unique experience. Flowering in the Himalaya begins with the advent of spring. But in the late summer, from July to September, rains in the Himalaya cover the region with the most exquisite flowers. Primulas in Sikkim are over by the end of April but in Himachal Pradesh, they bloom in May. This is also the time to photograph the flamboyant rhododendrons in the Himalaya.

While watching flowers on the hills, it becomes difficult to have a close look at those flowers which grow on inaccessible rock faces and cliffs. Here you may need a good pair of binoculars to identify the flowers. Binoculars certainly help in close observation of those characters of a plant which are not noticeable from a distance. The criterion to determine whether a pair of binoculars is suitable for watching flowers is the minimum focal distance of the binoculars. Ideally, you should find binoculars that can focus down to less than 2 m, otherwise you have to keep backing off to view the flowers. The closer you go, the better are the details.

Binoculars with a minimum focal distance greater than about 4 m should be avoided. Other features that need to be considered while buying a pair of binoculars are power, size, weight, and ease of focusing. Two numbers (e.g. 8 x 40) describe the basic features of the binoculars. The first number is the 'power', which means eight-power binoculars will make an object 80 metres away appear as large as if it was 10 metres away. The second number is the diameter of the objective lens. The larger the number, the brighter the image will be.

Besides being plain fun, watching flowers could be rewarding in many ways. It is an incentive to be outdoors, and can be challenging as well as gratifying as you learn to identify flowers, especially some new ones which you have never seen before. More importantly, like some birds and animals, flowering plants could be monitored as indicators of the health of a habitat, according to their presence or absence. In this way, some sensitive species could serve as an important tool in environmental protection.

It is indeed natural for a flower watcher to metamorphose into an avid flower photographer. The next chapter guides you in photographing flowers.

PHOTOGRAPHING WILD FLOWERS

Flowers have always been a photographer's favourite and it is indeed natural for the photographer to be attracted by the colours, shapes and forms of flowers. Flowers as subjects for photography could be approached both artistically as well as scientifically.

Fortunately today, photography has become less complicated with user friendly cameras and a variety of films and accessories. Flowers can be photographed with almost any type of camera. One often begins with the point and shoot camera. These cameras are usually good enough for scenic photography and for photographing flowers as big as a sunflower. It is much simpler to take a good colour photograph of a large, brightly coloured rose or dahlia. However, problems arise when it comes to photographing smaller flowers, which are spread out on tall spikes, or those which merge with their background. The most suitable camera for flower photography is the 35 mm single lens reflex camera (SLR), which is highly versatile, and simple to work with in the field. Moreover, user friendly SLR camera technology is at its peak today.

With the interchangeable lens system of the SLR, there is a good option to use lenses of different focal lengths. Flowers could be photographed using different lenses ranging from wide-angle lens to a long telephoto lens. A wide-angle lens can be used when the flowers are to be framed along with the habitat. A long telephoto lens enables you to photograph inaccessible flowers and also for the isolation of a flower from its background.

Much of flower photography involves close-up photography. The most suitable lens for this type of photography is the macro lens. These lenses are especially designed to give high quality close-up images. Macro lenses come in different specifications; 50 mm, 55 mm, 90 mm, 100 mm, 105 mm and 200 mm. A true macro lens is capable of giving a 1:1 image. The ratio 1:1 means that the image on the film is equal to the object at the minimum focusing distance. Of course, macro lenses are expensive, but when finances permit, do buy a macro lens. A macro lens in the 90-105 mm range is more suitable than 50-55 mm.

These special lenses being expensive and often scarce, nature photographers often resort to cheaper alternatives, though the results may not be good enough. Several telezoom lenses have macro facility which can be good enough for large to medium sized flowers. Supplementary close-up lenses that come in +1, +2 and +3 grades can be screwed on to the front of the standard lens for macrophotography. However, since the working distance between the lens and the subject is much reduced, the use of a flash often becomes impractical. And if the available natural light is inadequate, pictures tend to become soft and lack depth. But to start with, close-up lenses can be good. Attaching a 2x converter between the camera body and lens 'costs' a loss of two stops, and there is a small, but acceptable, loss of picture quality. The 2x converter's results could be bearable if the wind is light or absent and there is enough bright light. If the available natural light is too low, especially under forest cover or in an overcast situation, where hand-held photography is almost impossible, the use of a flash is often indispensable. The flash helps in getting sharp and brilliant pictures in most conditions. The rapid burst of light eliminates camera shake and freezes to some extent the swaying flowers in a windy situation. It is more useful when the available light is not sufficient to permit the use of small apertures like $f16$ or $f22$ to achieve maximum depth of field. The dark background created by the flash, though often disliked, in fact gives the appearance of sharpness because of a well-defined edge. Ideally, a flash that can be mounted on the camera is good enough to start with. Modern flashes have a dedicated system that works with the camera applying through-the-lens (TTL) metering, which is versatile in any situation. Such flashes now have controlled output and can be used 'invisibly' without causing harsh shadows. Better still is the ring flash, which is good for close-up work and gives almost shadowless results.

Fortunately, unlike other living subjects, flowers do not flee or fly away. In spite of this, your pictures may look fuzzy. This is not because the flower moved, but because the camera shook while you clicked. To avoid this, a cable release could be attached to a tripod-mounted camera. Though your touch may be gentle enough, it will not match the smooth squeeze of a cable release. Otherwise, for any exposure under 1/125th of a second, the camera should rest on something stable. For a low ground-level shot, a beanbag is most handy, but even a large stone can be a good support. For higher angles, a tripod is essential and often there is enough time to install it. The tripod should be the solid, sturdy type that can get right down to ground level.

Time exposure in low light is not recommended for beginners, as the result can be a total washout, or have a colour cast, though this can be corrected with a warm filter, or specific film, or under exposure. One can, however, master this too with trial and error. Photographing swaying flowers on a windy day is an impossible task. Even a gentle breeze can make your pictures fuzzy. Starting early in the morning by sunrise can be the most productive. Mornings are less breezy and without the thermals that are caused when the sun climbs high. As a remedy, high-speed film and faster shutter speed can help to some extent.

Photographing in the midst of a downpour should be avoided to protect your equipment from rain. But if you have to photograph in such a situation, a clear plastic umbrella can get you some memorable pictures. Similarly, to cut out breeze, a windshield can be improvised with a 4 x 1m transparent plastic sheet, 4 - 6 aluminium tent poles, metal tent pegs and a hammer. The plastic sheet is then arranged with the support of the tent poles and tent pegs in an arc of at least 240° around the plant. The shield may not be effective if the wind is strong.

Photographing flowers in natural light gives the best results and can be used very creatively. Overcast situations provide pleasantly even light that makes the flower colour come out more vividly. Light coloured flowers could be photographed against a light source. Spiny or hairy stems with delicate structures can be highlighted when photographed against the light. Back lit photographs can be very dramatic. In this situation, a lens hood is a must to reduce flares caused by direct light entering the lens.

Today's photographers are indeed blessed in having an array of films at their disposal, for specific results and to suit varied situations. Print or slide film? This question you should answer yourself. If you want to frame your pictures or are planning an exhibition or want your pictures to be displayed in a competition exhibition, then the obvious choice is print film. If you are a stock photographer or want to go commercial, then slide film (colour transparencies) is your natural choice. Photostock and advertising agencies prefer slides, as they are easy to store, their reproduction is of high fidelity, therefore preferred for calendars, greeting cards and magazines. For a beginner, print film is highly forgiving of mistakes of under- or overexposure, but slide film will show exactly where you went wrong.

But then, much of photography is learnt after making mistakes. One has to analyse and correct mistakes, in order to master this art. In addition, the most essential ingredient in flower photography is lots of patience to cope with changing light, or breeze which keeps your subject swaying. Be prepared to come the next day or the next year if you miss the season.

Useful Tips

1. Do not shoot the first flower that you come across. Usually there will be a better one. Once you have chosen your flower, look for the angle, attractive viewpoints, single flower or clusters. Wherever possible, choose young flowers.

2. Look out for flowers with a clear background. Of course, uniform green foliage makes the best out-of-focus background. A low angle shot against the sky or a higher angle against water can give a dramatically clear background. Even a dark bare rock can be a good backdrop. To isolate from a cluttered background, a medium to long tele or zoom lens is a perfect tool.

3. Check the flower carefully. A flaw in a petal or leaf, however small, may not make a pretty picture. Battered leaves or missing petals do not make commercially pretty pictures.

4. Move quickly, clouds can suddenly appear and attractive cross lighting can disappear in minutes. When there is wind, stay alert to click, as there are always momentary lulls.

5. Look carefully and look again in the frame of the viewfinder: compose what is essential and eliminate what distracts, both in the background and foreground. Include the background only if it enhances the composition. Before pressing the shutter, check the edges and the corners of the frame to ensure that everything is included in the composition.

6. Try out different angles since you have enough time with the subject not moving away. Frame carefully to show the flower to its best advantage. Try to avoid straight 'mug' shots. However, for identification of plants, straight shots are more practical than 'creative' shots. It is best to frame creeping plants directly from overhead, whereas erect tall plants could be shot from a low angle. For such low angle shots, it is best to mount the camera on a tabletop, tripod or on a ground spike.

7. Light coloured flowers could be isolated against the light source, but in a similar frame, a dark flower might lose detail, unless a fill-in flash is used.

8. Always click more than one frame of promising subjects, bracketing them with slight underexposure and then slight overexposure to get the best pictures. Underexposure deepens colours and darkens the background, while overexposure produces a pastel effect. Always bear in mind that film is the least expensive part of photography. Never hesitate to shoot a few extra frames or to shoot from different angles, especially those of a rare flower or a promising shot. Some classic photographic opportunities come only once in a life time.

9. In close-ups, everything cannot be sharp, therefore focus carefully and purposefully, and be sure that essential regions are more or less in one parallel plane to the lens. This way most of the subject will be in sharp focus. Use the highest f-stops (smallest apertures) for more depth, and if light permits, click at a reasonable shutter speed to avoid camera shake. If necessary, add light with a reflector or a flash.

10. Stop down the aperture only as much as you have to. If $f11$ is enough, then avoid $f16$ or $f22$. The more you stop down, the more cluttered the background and slower the shutter speed.

11. Try to shoot when the sun is still low and therefore warm. Slightly warm shots (slight golden cast) catch the eye, whether of competition judges, photo editors or fellow photographers. Cold shots (slight bluish cast) don't.

12. Finally, it is not only unethical but against the law to cut or collect flowers in protected areas like National Parks and Sanctuaries. Also, watch out where you put your feet, you may trample many flowers to get one.

CONSERVATION ALTERNATIVES

In terms of floral biodiversity, India has a rich flora of about 45,000 species ranging from algae, fungi, mosses, ferns, lichens, and conifers to flowering plants. Flowering plants make about 17,500 species and out of these, there are 4,900 species of plants endemic to the country. Many plant species are rapidly losing ground against the onslaught of man's exploitation and encroachment. So far, 19 plant species have been reported to be extinct. Another 41 species might be added to this list. 152 plant species have been listed as endangered, 102 are vulnerable and 251 rare among the Indian flora in the *Red Data Book* published by the International Union for Conservation of Nature and Natural Resources (IUCN). According to the IUCN's *Red List of Threatened Plants,* 12.5 percent of the world's flora is facing extinction.

Perennial Problem

A major factor contributing to the decline of plant populations is habitat loss due to human activities and unplanned development that has taken its toll. Such activities are mainly urban-industrial development, and quarrying, mining, hydroelectric projects, grandiose irrigation schemes which failed, timber extraction, tourism infrastructure and several others, which are primarily oriented towards meeting the demands of the urban and industrial sectors.

Large tracts of forest have been depleted of the original canopy for timber and firewood, much beyond their natural reparative power. Whatever natural vegetation was left over, herds of cattle and other livestock degraded the land further. Repeated intensive grazing prevents natural regeneration. While trees are cut for firewood, smaller branches are lopped off to feed livestock. It is well known that though it was geological events in the past that caused desertification, man and his grazing animals have accelerated the process. Even in the grasslands of the plains, shola-grassland of the Western Ghats and alpine meadows of the Himalaya, overgrazing has edged out several plant species. Overgrazing finally leads to the disappearance of palatable grasses. Unpalatable, distasteful plants dominate the landscape, making the area unfit for grazing. With the vegetation severely degraded, the climate too becomes increasingly arid and the land loses its agricultural potential. Only if the land-use is sustainable in a healthy ecosystem, can man reap a rich harvest.

On the hills, slash-and-burn shifting cultivation has been practiced traditionally, especially by forest dwellers. This type of shifting cultivation or *jhum,* as it is known in the northeast, was once a sustainable practice, but not any more. Earlier this practice was sustainable, as after the cleared patch was cultivated for some years, tribals moved further clearing a new patch, while leaving the earlier patch to regenerate. Now, however, with increased human population, such a practice is not sustainable as the cleared patch does not get enough time to regenerate and at the same time it is too degraded to yield good crops. This often leads to bringing more forest land under cultivation. Such land users need to be convinced to practice settled agriculture. Already the Government of India has a provision of special funds for settling *jhum* farmers. Best results could be achieved with non-governmental organisations and other local bodies working together to pursue positive communication between government officials and land users.

There are excellent government schemes available to guide and support land users to improve their yield. However, only the more vociferous and politically active land users get the benefits, while other remotely placed land users, unaware of such schemes, continue to till their degraded land and thus degrade more land to make up for the poor yield. Working on some of these problems, the Bombay Natural History Society during its Conservation Education Project, initiated in 1993, was able to liaise with and convince such land users, with government incentives, to go for settled agriculture, high yielding cattle breeds, stall-feeding of cattle and using fuel-efficient *chulhas* (stoves), which consume 40% less fuel-wood. While working on such problems, it is necessary to recognise the inextricable dependence and relationship of traditional communities with many of the ecosystems. Therefore, alternatives are to be provided for their needs and activities which cause irreversible damage to the habitats. Simultaneously, local populations should be encouraged to participate in eco-development, and to share benefits, thereby making them partners in conservation and sustainable users of natural resources. Assessing the success of such and similar projects implemented by other NGOs, there seems to be great potential in NGOs coming together to help the government for the welfare of the people and to conserve natural resources.

Medicinal Plants

Over 7,500 species of medicinal plants are estimated to be in use in India according to a survey conducted by the Ministry of Environment and Forests, Government of India*. There exists a thriving market for medicinal plants. For example, roots of the checkered orchid *Vanda tessellata* are sold at the rate of Rs. 200 per kg and the roots of a Himalayan orchid *Dactylorhiza hatagirea* are sold at Rs. 500-600 per kg. Tribals collect rhizomes of the endangered glory lily (*Gloriosa superba*) and sell it to middlemen at Rs.12 per kg. To meet the ever-growing demand, medicinal plants are usually collected from the wild stock. Records indicate that as much as 95 percent of raw materials required by pharmaceutical and drug manufacturers are collected from the wild. Most of them are collected using non-sustainable, destructive methods like collecting the entire plant, rhizome, tuber, roots or other reproductive parts like fruits and seeds. Such destructive collection methods are the major factors influencing plant populations. Further, low regeneration rate and loss of habitat add to the serious threat to medicinal plants. Such rapid depletion of medicinal plants from the wild requires urgent conservation measures.

Foreseeing the fate of some commercially important plants, efforts have been initiated to cultivate species like the Glory Lily (*Gloriosa superba*) which is endangered in the wild due to over collection. In the case of this species, multiplying plants by seeds was more effective than by tubers. Large scale commercial

---

* Amruth, Oct. 1997.

18

plantations have produced as much as 200 tonnes of seeds per annum*. On similar lines, plantations of *Coleus forskohlii, Passiflora incarnata* and *Mappia foetida* have been successfully undertaken.

One has to take into consideration that a commercial need for such resources is actually the need of the people and unless alternatives are provided through cultivation, wild species will not be secure. Already expertise, agro-technology and seed material is available in the country. The Ministry of Environment and Forests, Government of India, in collaboration with governmental as well as non-governmental organisations, has initiated projects for the conservation of wild medicinal plants and this has led to the establishment of a network of Conservation Sites, Conservation Parks and Development Sites.

Several drug manufacturing companies have started cultivating medicinal plants for their own requirement. This is a good beginning and will ensure a supply of genuine raw materials. Among the corporate-backed efforts in the conservation of endemic rare plants, the work of the Naoroji Godrej Centre for Plant Research (NGCPR) near Puné (Satara district) is indeed commendable. NGCPR is engaged in the conservation and cultivation of rare, endangered and endemic plants, besides the cultivation, improvement and mass propagation of medicinal and aromatic plants. Here Dalzell's Frerea (*Frerea indica*), one of the world's twelve most endangered plants, has been successfully cultivated, and efforts are underway to rehabilitate the plant in the wild, where it is threatened with extinction. NGCPR is engaged in several research projects aimed at conservation assessment and management plans for threatened endemic plants in collaboration with other NGOs, the University of Puné, and the Ministry of Environment and Forests, Government of India.

While there is an urgent need to conserve the country's rich floral biodiversity, it is also essential to recognise the vital role of plants in catering to the needs of the people as well as those of pharmaceutical and other industries. Besides medicine, herbal products range from health food, health drinks and cosmetics to personal care items. The international market for herbal products is growing fast and some multinational companies dealing in herbal products have an annual turnover up to $700 million**. Several such companies have been regularly importing Indian herbs on a large scale. Therefore, efforts should be taken to develop propagation techniques for large-scale cultivation of commercially important plants, to reduce the pressure on wild plants. This calls for a national policy framework for the conservation and sustainable utilisation of plants.

Innocent Marauders

With more and more wild places being opened up for tourism, orchids have been the most affected. Every tourist wants to take home these exotic beauties, little knowing that it is not only illegal to buy or remove wild orchids, but it is impossible

---

* *Hornbill*, Dec. 1998   ** *Economic Times*, 5th July, 1999.

to grow such specialised plants in a polluted urban environment. Often, such demand presses the locals to collect plants from the wild. Awareness among tourists with special informative posters can certainly help to check this vandalism.

When on botanical excursions, students in their misguided enthusiasm collect bag fulls of plant specimens. A majority of the material, collected in plastic bags, ends up as a soggy mass unfit for identification. Such wasteful activity needs to be regulated or even curbed. Often, batches after batches of students come to collect certain plant groups that are not common. This has already resulted in the disappearance of several plant populations. More than students, it is the teaching fraternity that needs to be reoriented in such practices. Backed by vigorous conservation awareness programmes, students should be encouraged: Not to pick! Not to uproot! The time has come to study plant communities as living ecological components and not just as dried herbarium specimens. There is a need for educational programmes on the importance of plants and their role as essential components of the country's biodiversity.

Vanishing Hedgerows

Hedgerows which border fields and roads enclosing properties used to be an essential feature in the Indian villages and even in semi-urban towns. But hedgerows are now fast disappearing. Property owners now seem to prefer 'clean' concrete walls or chain-link fences. It appears that hedgerows have now become an unnecessary clutter. In a survey done in Kerala, more than 800 species of plants were found in rich hedgerow compositions. They range from common lantana, clerodendrums, bamboo, barleria, eranthemums, ixora, wild gourds, morning glories, lilies, euphorbias, crotalaria, several grasses and ferns. Moreover, an astounding variety of insects, amphibians, reptiles, birds and small mammals were found in these miniature woodlands. Earlier thought to be unproductive clumps, it is now proved that hedgerows support a host of predators that feed on insect pests in farmlands. The beneficial residents of the hedgerow range from bees, wasps, mantids, spiders, frogs, lizards, snakes, shrews, mongooses, babblers, bulbuls, flycatchers, mynas, tailorbird, sunbirds, magpie-robin and several others whose main diet is insects. Since many of the birds nest in hedgerows, their insect consumption increases manifold after their young hatch. With so many insect-eating residents in hedgerows, the need to use pesticides is also reduced. Thus hedgerows not only support diverse plant life, but also keep the farms and orchards healthy.

Future of Mankind

Forests hold the future of mankind, because it is forests which bear the treasured gene pools of all cultivated plants in the form of their wild relatives. From time to time, many of these cultivated plants are threatened by diseases, which they are unable to fend off, against which their wild relatives have a natural immunity. This immunity is used by plant scientists to strengthen cultivated varieties by selective breeding, so that there is enough food to feed the growing populace. Besides, there may be several others, which could be the sources of new drugs, medicines or other useful biochemicals.

One should not forget that India is considered to be one of the world's 12 centres of the origin of cultivated plants. India's biodiversity has contributed 51 species of cereals and millets (grains); 104 species of fruit; 27 species of spices and condiments; 55 species of vegetables and pulses; 24 species of fibre crops; 12 species of oil seeds and various wild strains of tea, coffee, tobacco and sugarcane.

There are several more plant species waiting to be discovered, but they may disappear even before we come to know about their existence. This will be a permanent loss to mankind.

Sacred Groves

Today, we have legally protected areas like national parks and sanctuaries. In ancient India, almost every village had a sacrosanct patch of forest dedicated to a deity, which was left untouched by villagers. Some of these Sacred Groves or temple forests still exist as pockets that hold ancient forests, and therefore are rich in biodiversity. Even today, these ancient forest patches are protected by the religious sentiments of the people and are found throughout India. This fact is now well recognised even by the Ministry of Environment and Forests, which has already initiated the process of documentation of the Sacred Groves all over the country, which possibly hold some of the last treasures of biodiversity. The Ministry is co-ordinating this documentation with several governmental and non-governmental institutions. The plan is to document temple forests all over the country, assess their biodiversity and finally assist people in guarding their own treasures.

Unified Effort

There is need for a unified effort to achieve results. Under the co-ordination of the Ministry of Environment and Forests, concerned NGOs, government institutions, forest departments, botanical gardens, plant collectors, traders, botanists, nurserymen, conservationists and naturalists should be brought together under one umbrella to form an Interest Groups Network. These interest groups can then lobby for proper legislation or amendment of the existing laws to regulate internal and external trade and promote funding to protect, study and monitor threatened species of plants and their habitats, identify areas of high plant diversity and ensure their conservation.

FAMILIES OF FLOWERING PLANTS

A brief introduction to the families of flowering plants represented in the book is given below:

Buttercup Family   **Ranunculaceae**

A large mixed group consisting mostly of herbs and some climbers that mainly grow in cool climates of hills and mountain ranges like the clematis which represents this family. Several plants of this group like anemones (*Anemone*), marsh marigold (*Caltha*) and clematis (*Clematis*) occur in the temperate Himalaya. Mostly perennial, these plants have primitive flowers with many separate parts spirally arranged and divided leaves. Some plants of this group like anemones, buttercups and clematis are poisonous, while some are used in medicine. Family name derived from Latin *rana* (a frog) as several species grow in wet places.

Waterlily Family   **Nymphaeaceae**

Waterlilies (*Nymphaea*) are a group of floating aquatic herbs closely related to the lotus (*Nelumbo*). They have large underwater creeping stems (rhizome). Waterlilies have heart-shaped leaves, with a deep notch, that usually float over the water surface. Waterlilies are popularly cultivated in ponds. Genus *Nymphaea* is named after Nymphe, a water nymph of Greek mythology.

Lotus Family   **Nelumbonaceae**

Lotus is a well known plant. Lotus leaves are round, entire without any deep notch and often emerge out of the water. Flowers are both rosy pink and white. Lotus is cultivated for its ornamental flowers in lakes and temple ponds. Rhizomes, fruit and leaves are eaten as vegetables. Flowers used in making perfume. Lotus honey, flowers and rhizomes are used in traditional medicine. Generic name *Nelumbo* is from Sinhalese name of the plant.

Poppy Family   **Papaveraceae**

This family has plants that are ready colonisers of vacant lots. The best example is the Mexican poppy (*Argemone mexicana*), a native of Central America that has spread throughout the warm countries in the world. Other members of this group like Himalayan poppy (*Meconopsis*) and *Corydalis* sp. prefer the cool temperate climate of the hills. This group has both annuals as well as herbaceous perennials, some are poisonous. The opium poppy (*Papaver somniferum*) is used in making pain-killing medicines as well as narcotic drugs.

Caper Family   **Capparaceae**

This is a mixed group of herbs, shrubs and trees. Capers are known for their backward curving thorns and no trekker has escaped their clutches. The sacred barna (*Crateva magna*) is a beautiful sight when it blooms after shedding leaves. Flowers are typical with long numerous stamens. Butterflies belonging to the White and Yellow group (Pierids) prefer to lay eggs on plants of this group.

Cleome Family   Cleomaceae

The scented, pink flowering spider flower (*Cleome speciosa*) is popularly grown in gardens from where it has escaped to grow wild. Flowers are typical with several long stamens. Both Common Spider Flower and Yellow Spider Flower have spread quite aggressively in recent years. Some species of this group have an unpleasant odour. This group of plants had been earlier placed in Caper family.

Purselane Family   Portulacaceae

Succulent, sun-loving, low growing herbs, among which the common purselane is a common intruder in gardens. Several colourful varieties of the sun-loving Chinese rose and office time are popularly planted on windowsills and in garden beds. Common purselane is often sold as a pot herb. Short-lived flowers remain open during the hottest part of the day. Common purselane (*Portulaca oleracea*) is a favoured food plant of the Eggfly butterflies (*Hypolimnas* spp).

Mallow Family   Malvaceae

A well-known cosmopolitan group of widely distributed herbs, shrubs and trees. The ubiquitous red shoeflower (*Hibiscus rosa-sinensis*) is the commonest among the mallows. Ladies'fingers or bhendi (*Abelmoschus esculentus*) is a popular vegetable, while the bhendi tree (*Thespesia populnea*) is commonly planted near habitations and roadsides. Several ornamental varieties of Hibiscus are grown in gardens for their large attractive flowers.

Sterculia Family   Sterculiaceae

Except for a few herbs, this family has mainly shrubs and trees. The white bark of the ghost tree (*Sterculia urens*) makes this deciduous tree conspicuous in the forest, especially on a moonlit night. The profuse flaming flowers of coloured sterculia (*Firmiana colorata*) stand out among the leafless forests in summer. East Indian screw tree (*Helicteres isora*) never fails to attract birds for nectar. The scarlet Phoenicean (*Pentapetes phoenicea*), a native of Bengal is popularly grown in gardens in the northeast and is often seen growing on garbage dumps in cities.

Jute Family   Tiliaceae

A small, but important family of herbs, shrubs and trees. *Phalsa* tree (*Grewia asiatica*) is known for the cooling *phalsa sherbet* made from its fruit. The well known jute fibre is mainly obtained from *Corchorus capsularis* and *C. olitorius*. Among wild flowers, this family is represented by forest burbush (*Triumfetta rhomboidea*), an annual whose flowers open only in the late afternoon.

Linseed Family   Linaceae

Linseed plant or *alsi* (*Linum usitatissimum*) is cultivated in India mainly for linseed oil. Though its native country is unknown, it is supposed to have originated from Egypt, where it has been cultivated since the time of the Pharaohs. It is also popularly called flax-plant. Here it is represented by the dainty little Mysore linseed, commonly seen on the hills as the monsoon trails off.

Caltrop Family   **Zygophyllaceae**

This group consists mainly of shrubs and a few herbs, with most having leaflets arranged in two rows on either side of the stalk. A typical example of this family is the Puncture Plant or Land-caltrop (*Tribulus terrestris*). Plants of this group are found in tropical, sub-tropical as well as temperate regions. Fruits have spines, hence the common name Caltrop, which was the four-pointed weapon used by gladiators.

Geranium Family   **Geraniaceae**

Geraniums have always been popular garden flowers, especially in the hills. This is a family of annual and perennial herbs and a few small shrubs that prefer temperate and sub-tropical regions. The African geranium (*Monsonia senegalensis*), which occurs in the drier regions of the country represents this group in this book. Geranium oils are used in cosmetics.

Wood-Sorrel Family   **Oxalidaceae**

A small group of annual as well as perennial herbs that usually go unnoticed in untended gardens and backyards. Some like broad-leaved sorrel (*Oxalis latifolia*) are stemless and bulbous. Common sorrel (*Oxalis corniculata*) is a common intruder in lawns and among potted plants. Except for the little tree plant (*Biophytum sp.*), whose leaves are slightly sensitive, all wood-sorrels (*Oxalis* sp.) have three leaflets. Some wood-sorrels (*Oxalis* spp.) are known to be toxic to animals. Genus name *Oxalis* is derived from Greek *oxys* (acid), referring to the presence of sour oxalic acid.

Balsam Family   **Balsaminaceae**

A popular group of succulent herbs, mainly annual and some perennial. These brittle-stemmed herbs are known for their attractive flowers recognised by a 'spur' on the lower large pouched sepal. They are usually seen along the streams, wet rocks and wet grassy slopes in tropical forested regions as well as in temperate Himalaya. Balsams have always been favourite annuals for garden beds. Several ornamental varieties with double flowers have been developed. Ripe seed bearing capsules 'explode' scattering out the seeds. Balsams are used in cosmetics and they have anti-bacterial properties. Flowers preferred by butterflies, moths and bees.

Vine Family   **Vitaceae**

This is the family to which the grape vine (*Vitis vinifera*) belongs. Plants in this family are mainly woody climbers with a few shrubs. Typical vines, several of these climbers have tendrils for support. Fruits are typical grape-like fleshy berries. Several species of hawkmoths lay eggs on some of these plants.

Pea Family   **Papilionoideae**

A large group that includes an assortment of wild as well as cultivated species of peas and beans of which some are important food crops. Flowers in this family have five petals. One upper exterior petal or standard forms a hood. Two inner petals on either side form a pair of wings. Two other lower petals parallel to each other form a beak-shaped keel. The pod is called legume and contains one or more seeds.

Peacock Flower Family   Caesalpinioideae
Orange and yellow flowering shrubs of peacock flowers (*Caesalpinia pulcherrima*)
and *gulmohur* (*Delonix regia*) are some of the more prominent members among
these legumes. There are also others like tamarind (*Tamarindus indica*), copperpod
tree (*Peltophorum pterocarpum*), *Bauhinia* and a number of *Cassia* species in
this family. The family is named after Andreas Caesalpini (1519-1603), an Italian
botanist.

Touch-me-not Family   Mimosoideae
An interesting group having touch-me-not (*Mimosa pudica*), thorny *Acacia* spp.,
rain tree (*Samanea saman*), sirish (*Albizia lebbeck*), painted thornbush
(*Dichrostachys cinerea*), subabul (*Leucaena leucocephala*) and mesquite or *ganda
baval* (*Prosopis juliflora*). Many of these have flower heads like powder puffs. A
woody climber called the giant rattle (*Entada rheedii*) that has large woody pods
is also a member of this group. Sensitive neptunia (*Neptunia oleracea*), a floating
aquatic is an odd member of this family.

Terminalia Family   Combretaceae
This family is known for the large trees like arjun (*Terminalia arjuna*), beheda (*T.
bellirica*), Indian almond (*T. catappa*), *T. chebula* and ain (*T. crenulata*). Another
popular member of this group is the ornamental Rangoon creeper (*Quisqualis
indica*), that bears bunches of fragrant pink and white flowers.

Blackmouth Family   Melastomataceae
This family has trees, shrubs and herbs like *Osbeckia* spp., *Memecylon* spp. and
Melastoma spp. *Osbeckia* and *Melastoma* have very similar showy flowers, except
flowers of the Malabar blackmouth (*Melastoma malabathricum*) which have yellow
and reddish-purple anthers alternately. The edible pulp of the melastoma fruit
stains black, hence the generic name *Melastoma*, which means "blackmouth" in
Greek. *Osbeckia* and *Melastoma* are usually seen along forest streams.

Henna Family   Lythraceae
A small group of trees, shrubs and herbs, which is also called the Loosestrife
family. Trees like queen's flower (*Lagerstroemia reginae*) and other *Lagerstroemia*
spp. are well-known. Shrubs like fire bush (*Woodfordia fruticosa*) and henna
(*Lawsonia inermis*) are commonly seen. Since ancient times henna or *mehndi* has
been used in comestics, and today with cosmetics going herbal internationally,
henna is much in demand. *Lagerstroemia* spp. are preferred food plants of the
Atlas moth, while the Moon moth lays eggs on henna too.

Water-Chestnut Family   Trapaceae
A family of floating aquatic plants that prefer still waters. The water-chestnut is
cultivated for its edible fruit that is eaten raw when green, cooked when mature and
black and made into flour. Family name is derived from Latin *calcitrappa* (caltrop,
a four-pointed weapon) referring to the horned fruit.

Passion Flower Family   **Passifloraceae**

A group of vines with tendrils to climb like Krishna kamal (*Passiflora edulis*) and (*P. foetida*) that are native to tropical South America and have now totally naturalised in Asia and Africa. There are some endemic species too, but being widely planted in gardens and cultivated for fruits, the exotics are better known. Besides man, birds also cause dispersal of the seeds. Often groups of black caterpillars with soft fleshy spines could be seen on the undersides of the leaves of these vines. These soon pupate on the vines and Tawny Coster butterflies emerge. Family name derived from Latin *passio* (passion) and *flos* (a flower).

Cucumber Family   **Cucurbitaceae**

A family of herbaceous climbing or trailing plants with tendrils like the cucumbers, melons and gourds. Flowers are yellow or white with five petals. Each flower is either male or female and in some species male and female flowers grow on different plants. This is an important family, with several wild parent stocks of the present cultivated varieties of cucumbers, melons and gourds.

Begonia Family   **Begoniaceae**

A small family of herbs and shrubs, that are often popular as garden plants. The plants prefer sheltered moist conditions under shrubs, trees or rocks. Most have tubers or rhizomes. Flowers come in clusters. Male and female flowers are considerably different. In the wild, begonias are seen in moist forests and hill-stations. This family is named after Michael Begon (1638-1710), Governor of French Canada.

Shore Purselane Family   **Aizoaceae**

A family of low growing, spreading herbs with succulent fleshy or thin leaves. Here this group is represented by sand & stone plant (*Corbichonia decumbens*) and shore purselane (*Sesuvium portulacastrum*).

Coriander Family   **Umbelliferae (Apiaceae)**

Mainly herbs having small flowers arranged in umbrella-shaped flower heads called umbels. *Umbella* in Latin means a sunshade. This group contains several well-known important plants like carrot (*Daucus carota*), cumin or *zeera* (*Cuminum cyminum*), coriander (*Coriandrum sativum*) and *brahmi* (*Centella asiatica*) which are famous for flavouring food and making medicines.

Coffee Family   **Rubiaceae**

A large group mainly of shrubs, climbers, trees and some herbs. Some well-known plants in this group coffee are (*Coffea arabica*), *Mussaenda*, *Gardenia* and *Ixora*. *Gardenia* is a preferred food plant of the Bee Hawkmoth (*Cephondes hylas*), while the Commander butterfly lays eggs on *Mussaenda*.

Sunflower Family   **Compositae (Asteraceae)**

This is the largest family of flowering plants containing over 20,000 species growing all over the world. A daisy or a sunflower is a composite, which means that these flowers are made up of many small flowers called florets. The central flowers or disc florets are often tubular, while the outer ray florets are long, flap-like 'petals'.

Together, they function as a flower. Several species have seeds with a silky bunch of hairs called pappus, that helps in their dispersal by wind.

Sea Lavender Family   Plumbaginaceae
This group has herbs, shrubs and climbers that are well adapted to grow in coastal areas. The cape leadwort (*Plumbago auriculata*) is a popular garden plant that has sky-blue phlox-like flowers. Its white flowering cousin *P. zeylanica* grows wild. Some members of this family occur high in the Himalaya between 3,000-4,500 m. Family and genus name derived from Latin *plumbum* (lead).

Primrose Family   Primulaceae
Primrose flowers are associated with the arrival of spring in the Himalaya, where primulas or primroses bloom gregariously on open slopes and meadows. However, different species bloom in different months, ranging from April onwards. This is a mixed family of herbaceous plants. Himalayan primroses are often seen on leafless stems and leaves grow much larger once the flowering is over. The scarlet pimpernel (*Anagalis arvensis*) is poisonous to animals. However, it is now used in making eco-friendly insecticides.

Jasmine Family   Oleaceae
Also called the Olive family, this is a mixed group of shrubs, woody climbers and trees. Among this group, jasmine species are most prominent, known for their sweet fragrance. Several varieties are cultivated commercially on a large scale and popular as garden ornamentals. Other known members are Olive (*Olea* spp.) and coral jasmine or *parijat* (*Nyctanthes arbor-tristis*), whose fragrant, orange-stalked white flowers are seen scattered in the mornings below the shrub.

Oleander Family   Apocynaceae
A familiar group having trees, shrubs, vines, woody climbers and some herbs, which usually have poisonous milky sap. Flowers are often showy and fragrant. Dark blue berries of Christ's thorn or *Karaunda* (*Carissa congesta*) are often sold by tribals in summer. Among the garden favourites that belong to this group are golden yellow flowering trumpet vine (*Allamanda cathartica*), pink and white flowering oleander (*Nerium oleander*), yellow oleander (*Thevetia peruviana*) and the frangipani or pagoda tree (*Plumeria rubra*). There are several plants of medicinal value like easter tree (*Holarrhena pubescens*), serpent root (*Rauvolfia serpentina*), periwinkle (*Vinca* spp.) and (*Wrightia tinctoria*).

Milkweed Family   Asclepiadaceae
An interesting family of herbs, shrubs and woody climbers, most having a distasteful milky sap; hence the name. Fruits are usually in pairs and in some species, they are curved and pointed like a bull's horns. Flattened seeds have a tuft of silky hairs which help in their dispersal by wind. The most common among the milkweeds is the giant milkweed (*Calotropis gigantea*). Some like the flytrap vine (*Ceropegia* sp.) and Dalzell's Frerea (*Frerea indica*) are endangered. Milkweed butterflies like the Tigers and Crows lay eggs on several of the milkweed plants. The scientific

name of this family is derived from Greek *Asklepios*, god of medicine, due to its medicinal properties.

Bogbean Family   **Menyanthaceae**
Aquatic herbs with underwater creeping and rooting stems, and floating leaves; thus they superficially resemble water lilies. The plants grow gregariously and at times can be very dominant covering the entire water surface.

Borage Family   **Boraginaceae**
In this group barring a few trees and shrubs, a majority are herbs. Plants are usually hairy and aromatic. The most familiar among this group is the Indian turnsole (*Heliotropium indicum*) and Indian borage (*Trichodesma indicum*). Another often seen is the *bokar* or *gunda* or cordia tree, whose fruits are pickled. Flowers of these herbs are a favourite with butterflies, bees and other insects. Milkweed butterflies are seen around these plants which yield an essential alkaloid that is necessary for the butterflies' reproduction.

Glory Family   **Convolvulaceae**
A popular group of vines, climbers, herbs and shrubs with a majority having typical funnel-shaped flowers. An oddity in the group is the dodder (*Cuscuta* sp.), a leafless parasitic twiner. Contrary to the usual belief, flowers of all glories do not open in the morning. While some *Ipomoea* sp. open just after sunrise, there are others like the tiger's paw glory (*Ipomoea pes-tigridis*) and night glory (*Rivea* sp.) that open around sunset. Sweet potato (*Ipomoea batatas*) is commercially cultivated for its tubers. Several glories are popular as ornamentals while those like woolly elephant climber (*Argyreia nervosa*) and little glory (*Evolvulus alsinoides*) have been used medicinally. Some members like the hedge glory (*Ipomoea carnea*) are distasteful to animals.

Nightshade Family   **Solanaceae**
An important family that includes potato (*Solanum tuberosum*), tomato (*Lycopersicon esculentum*), brinjal (*Solanum melongena*) and capsicum (*Capsicum annuum*). Tobacco too belongs to this group, which also has many poisonous plants like the *dhatura* (*Datura metel*) and several which have valuable medicinal properties. Most plants in this group are herbaceous.

Figwort Family   **Scrophulariaceae**
A group having a strange mixture of aquatic and marsh herbs like the *Bacopa* and *Limnophila* and root parasites like the purple witch (*Striga* sp.), while some like lousewort (*Pedicularis* sp.) and the dainty sopubia (*Sopubia delphinifolia*) are partial parasites. Family name is derived from Latin *scrofulae* (a swelling of the neck glands) referring to the medicinal properties of some plants in this family.

Broomrape Family   **Orobanchaceae**
A group of queer parasitic plants that lack green leaves, but instead have coloured, fleshy, scale-like leaves. The forest ghost flower, (*Aeginetia indica*) is a denizen of dark undergrowth of the forests where not much sunlight reaches. The plant is

parasitic on the roots of other plants like yams (*Dioscorea* sp.). The handsome yellow broomrape (*Orobanche cistula*) is known to parasitise the roots of *Salvadora persica* and sometimes *Calotropis* sp.

Sesame Family   **Pedaliaceae**
A small family of herbs and small shrubs with sticky sap and tubular bell-shaped flowers. It includes the economically important sesame or *til* (*Sesamum orientale*), cultivated as an oil crop. Sesame also occurs in the wild as an escape. The purple sesame (*S. laciniatum*) and common pedalium (*Pedalium murex*) are seen in the drier regions. The devil's claw (*Martynia annua*), a native of Mexico, is now naturalised in India.

Acanthus Family   **Acanthaceae**
The family consists mainly of shrubs, annuals or perennial herbs and some climbers. Flowers are mostly solitary, but some species bear them in small clusters. The flowers are often 2-lipped. *Acanthus* is derived from Greek *Akanthos* (a thorn) referring to the often spiny leaves and flower bracts. A typical acanthus is the sea holly (*Acanthus ilicifolius*) that has spiny leaves. Stringed saffron *Crossandra* flowers are often worn by ladies to adorn their hairdos. *Thunbergia grandiflora*, is commonly planted in gardens for the large pale-blue flowers. Members of this family are favoured food plants of the Pansy group of butterflies.

Vervain Family   **Verbenaceae**
A mixed family of herbs, shrubs, lianas and trees that range from the creeping low bank mat (*Phyla nodiflora*) to the mighty teak tree (*Tectona grandis*). Garden verbena never fail to give the garden bed a splash of colour. The most well-known member of this group is the common lantana (*Lantana camara*), the most successful among the alien invaders. Several varieties of *Clerodendrum* are popularly planted in gardens.

Mint Family   **Labiatae (Lamiaceae)**
This is the family of strong smelling herbs and shrubs. Well-known among them is *tulsi* (*Ocimum*), mint or *pudina* and the colourful *Coleus* in the garden. Most of these plants have flowers with petals that are joined together to form a tube, which usually has two lips. These aromatic plants are used directly for their strong smelling oils as food flavouring. This special aroma is exuded from the numerous glandular hairs that cover the plant.

Four o'Clock Family   **Nyctaginaceae**
Gulbaxi or the four o'clock (*Mirabilis jalapa*) is so called because its flowers open in the late afternoon and remain open during the night. Even the ubiquitous bougainvillea (*Bougainvillea glabra*) is from this family, which has several more trees, shrubs and herbs. Here *Boerhaavia diffusa* represents this family.

Cockscomb Family   **Amaranthaceae**
A well known group of herbs and shrubs, this large family also has some garden plants and pot herbs. Small flowers are in terminal or axillary clusters. Several of

the Amaranthus species are cultivated widely as pot herbs. Garden annuals like love-lies-bleeding (*Amaranthus caudatus*), Joseph's coat (*Amaranthus tricolor*), crested celosia (*Celosia cristata*) and bachelor's buttons (*Gomphrena globosa*) are quite popular. The scientific name of this family is derived from Greek *amarantos* (unfading), referring to the long-lasting flowers.

Buckwheat Family   Polygonaceae

This is a mixed group of herbs and shrubs. The common marsh buckwheat (*Polygonum glabra*) is typical of this group, having small flowers in dense clusters. Several members are common on the cool heights of the hills. The gregarious fleece flowers (*Bistorta* spp.) occur at 3,000-4,800 m in the Himalaya. Buckwheat (*Fagopyrum* sp.) is widely cultivated in the Himalaya for grain. Family name is derived from Greek *poly* (many) and *gony* (a knee) referring to the jointed stems.

Birthwort Family   Aristolochiaceae

A group of herbs and climbers. Flowers of this group of plants are indeed strange shaped and often smell foetid to attract flies that enter into the narrow flower tube and get trapped till pollination is affected, and the flower begins to collapse to let the trapped pollinators escape. Plants are poisonous and bitter, but are favourite foodplants of the red-bodied swallowtails like Common Rose and Crimson Rose butterflies and also the large Birdwing butterflies. Seeds come in an inverted umbrella-like capsule. Scientific name of the family is derived from Greek *aristo* (best) and *lochia* (childbirth), referring to supposed medicinal properties.

Daphne Family   Thymeliaceae

This group consists mainly of shrubs that occur both in the temperate as well as tropical regions. Here this group is represented by a very common shrub of the Western Ghats, the fish-poison bush (*Gnidia eriocephalus*), which flowers in the summer. The stem and leaves are locally used as fish poison; fish do not die, but are stunned and float along the bank.

Mistletoe Family   Loranthaceae

Clumps of drooping wiry branches of these parasitic plants are seen on trees on which they depend for support and nutrition. Though they have green leaves for photosynthesis, they are attached to the host plant by modified roots. Tubular flowers are tailor-made for nectar seeking birds, which pollinate the flowers. Ripe pulp of the berries is eaten by birds, while sticky seeds are rubbed off their beaks against the branch, thus helping in their dispersal. Butterflies like Common Jezebel and Gaudy Baron lay eggs on these plants.

Spurge Family   Euphorbiaceae

This large group has a mixed lot ranging from trees, vines to soft-wooded shrubs and herbs with some typical spurges having a distasteful, milky sap. Several spurges are familiar garden plants, like the winter flowering poinsettia (*Euphorbia pulcherrima*) and spiny crown of thorn (*E. milii*) and *Jatropha* sp. Useful and commercially important castor (*Ricinus communis*) and amla (*Phyllanthus emblica*) belong to this group. Several species of trees like *Mallotus*, *Macaranga* and

Bridelia are also in this family. Erect, cactus-like spiny euphorbias are often seen planted along the fields as a hedge. The milky sap of these plants is poisonous to animals.

Orchid Family   Orchidaceae

This is the second largest group of perennial flowering plants that grow on land and on trees. The tree-dwellers are non-parasitic (epiphyte) and have aerial roots that absorb moisture and nutrients. Some orchids are saprophytic, and feed on dead organic matter, their leaves are small, brownish and scale-like. Both ground and tree-dwelling orchids have swollen stems called pseudobulbs, where water and nutrients are stored. Orchid flowers are highly specialised to attract pollinators. Most have very small seeds, which require symbiotic fungi to germinate. The world over, there are about 18,000 species of orchids and are most valued by florists and orchid fanciers. *Vanilla planifolia*, a climbing orchid, is the source of the well-known vanilla essence.

Ginger Family   Zingiberaceae

These herbs have fleshy, branched underground rhizomes that are usually aromatic, with canna-like leaves. Flowers are borne either at the tips of the stems or on separate flowering shoots that grow next to the plant. In the wild, these herbs prefer forested hilly regions. The prominent members among this group are the turmeric or *haldi* (*Curcuma domestica*) and ginger (*Zingiber officinale*). Important spices like the dark large cardamom (*Amomum subulatum*) is cultivated on the shady forested slopes in Sikkim and eastern Nepal and the smaller white cardamom (*Elettaria cardamomum*) cultivated on the forested hills in south India also belong to this group. Family name is of ancient origin probably from pre-roman *srnga* (a horn) and *ver* (a root), from which Latin *zingiber* and English *ginger* have been derived.

Canna Family   Cannaceae

Though Indian shot (*Canna indica*) is probably the most commonly seen, several horticultural varieties have been developed which are commonly planted in gardens. Being robust and quick growing, canna is aggressive enough to colonise waterlogged areas. Also, it is often seen near habitations planted as screens or hedges. All species of canna came from tropical and subtropical America and have naturalised in Asia and Africa. Genus name is derived from Greek *kanna* (a reed).

Banana Family   Musaceae

Most familiar and economically important, Banana plants are native to tropical Africa, eastern Asia, Australia and the South Pacific. In India, wild banana is common on the hills in south India, eastern Himalaya and north eastern region. The banana plant is one of the largest herbaceous plants.

Daffodil Family   Amaryllidaceae

Members of this family are mainly perennial herbs having bulbs and tubers. This family is closely allied to the Lily family. The large flowered amaryllis lilies are ever popular with gardeners. The fireball lily (*Haemanthus multiflorus*) is also common in Indian gardens. Just like typical lilies, bulbs lie dormant underground till the next season. Scientific name takes after a shepherdess in Greek mythology.

Star Lily Family   **Hypoxidaceae**

A group of dainty herbs that sprout with the rains, and in the spring (in temperate region). Widespread in warm temperate to tropical region, these herbs are absent in Europe or Northern Asia. These small plants are sometimes classified as member of the Daffodil family Here it is represented by the Yellow Ground Star.

Lily Family   **Liliaceae**

Lilies are mainly perennial herbs that grow from underground stems (corms, bulbs, tubers or rhizomes) belonging to a very large, mixed group. The beautiful glory lily, aloes and asparagus are medicinally important plants from this group. Several varieties of lilies are popular as garden plants for their beautiful ornamental flowers.

Water Hyacinth Family   **Pontederiaceae**

This group is also known as Pickerel Weed family. Best known among this family is the water hyacinth (*Eichhornia crassipes*), which is one of the most troublesome plants, clogging waterways and other waterbodies. This group is made up of rooted or floating aquatic herbs. The slender spongy lesser water hyacinth (*Monchoria vaginalis*) is often seen along the rice fields and streams.

Spiderwort Family   **Commelinaceae**

A group of succulent herbs, mainly annuals, seen around human dwellings like the common garden commelina (*Commelina bengalensis*). Flowers are usually surrounded by a boat-shaped bract. Plants of this group are usually gregarious. Strangely, some like the common garden commelina bear underground flowers, beside their normal blue flowers. These underground flowers do not open, they self-pollinate and bear better seeds. Wandering Jew (*Zebrina pendula*) is a familiar garden plant often planted in hanging baskets. Cyanotis too are fairly common annuals seen during the rains.

Arum Family   **Araceae**

Members of this group are all herbs like taro, *suran* or elephant foot yam have long been cultivated by man and sold by vegetable vendors. Typical tall tapering flower almost enclosed in a green or coloured spathe. Both leaves and tubers of taro (*Colocasia esculenta*) are edible, whereas elephant foot yam (*Amorphophallus*) and *Arisaema* tubers are cooked. It is an essential requirement of this family that the plant or its parts have to be cooked with acidic tamarind to neutralise calcium oxalate in the plant that can cause an otherwise unforgettable itch in the throat.

GARDENING WITH WILD FLOWERS

Gardening with wild flowers is certainly challenging, but once successful, your garden will become a sanctuary for many other wild creatures like birds, lizards, frogs, butterflies, moths and other insects. If you live in the heart of a bustling city, you will be pleasantly surprised to watch the bees and the butterflies visiting your garden, even if it is on your windowsill. Of course, in the beginning, collection of seeds, cuttings and saplings will take time, but with some patience and experience you can soon master this unique pursuit.

Collecting seeds, bulbs, saplings and cuttings

This is a major exercise, which will involve a lot of plant hunting. In India, seeds of wild flowers are not readily available in nurseries and you will have to collect them yourself. Throughout this collecting process, one should always resist the temptation of collecting an endangered or rare plant from its habitat. No plant should be collected from a national park or a wildlife sanctuary. It is not only unethical, but illegal also.

Seeds come in all shapes and sizes — while some are as fine as dust, others are like stones or flakes. As the monsoon trails off, most annuals will have seeds ready for harvesting. On the other hand, the seeds of perennials ripen just before the rains. Store the seeds in a cool, dry place till you plant them.

For plants from the Sunflower family, look for seeds once the petals have withered and a seed head has formed. The seeds are ripe when the seed heads are brown. Cut the seed heads, seal them in a paper envelope and label it clearly. Similarly, seeds of glories, mallows and barlerias can be collected from half dried plants. Collect the pods of the Pea family when they are turning brown; some may snap open as soon as you touch them. While collecting balsam seeds, hold a paper envelope below the mature pods and press them slightly, the ripe ones will pop open to scatter the seeds. Most gourds will turn brown when mature, but some species turn scarlet-orange and pulpy. Dry gourds need to be shaken vigorously over a newspaper to collect the seeds, while seeds from pulpy gourds have to be separated and dried before storing. It is always safer to sun dry collected seeds for a week to avoid fungus and insect infestations.

As for bulbs and tubers, they should be collected during the first few weeks of the rains when they flower, so that you know what you are collecting. This is also the most appropriate time to collect and transplant young saplings, if you are able to recognise the species. For example, if you wish to collect karvi saplings, carefully dig out young plants during the monsoon after the karvi has flowered. Monsoon is also the right time to raise cuttings of perennials like the chaste tree or the lantana — apply the same rule of thumb as for collecting and planting other garden plants. Some perennials are difficult to grow from cuttings, while others like the capers and Christ's thorn have to be raised from seed. Much of this technique has to be learned by trial and error and one has to be as innovative as possible.

For water lilies, lotus and crested snowflakes, it is advisable to transplant the rootstock straight into the garden pond. Broken mature leaves of star water lilies will produce young plantlets if left undisturbed in water. Clumps of bankmat, bearded marsh-star, marsh carpet and other water loving plants could be directly transplanted around the pond.

Planning the garden

Since different plants flower in different seasons, the garden could be planned accordingly. The most important factor that governs the well-being of flowering plants is the amount of sunshine, and much depends on the location of the garden. Preferably, flowering plants should get at least four to six hours of sunshine. Gardens facing east and southeast will get the maximum sunshine.

While planting, taller plants should be planted at the back and smaller ones at the front of the bed. Allow plenty of room for moderate sized plants. Larger shrubs will require even wider beds. Anticipate the full grown size of the plant species you select, and plant them so that they don't crowd out each other. Leave space in between plants for air circulation and room for growth. Many plants do not grow tall and will not require constant pruning.

Natural rocks and pebbles could be effectively used to create a rockery. Rocks and pebbles help in holding back the soil and are used to make different levels. Planting could be done to give a cascading effect. You may consider yourself lucky if your garden has some natural rocky areas. Instead of digging the rocks out, design a rock garden around them. Similarly, a rock pool or a pond could be added to the garden for aquatic plants and water dwelling creatures.

Plant Selection

Finally, selection of plants depends on factors like the plant's maximum size and its adaptability. Be aware of its need for sun and water, and be sure to put it in an area of your garden where it will get what it needs. The tougher ones you will soon know, as they will do better.

Plant shrubs and perennial climbers first, as they take longer to reach maturity than the annuals do. They will require time to grow while you plant the less permanent annuals in your landscape. Needless to say, select the plants carefully. Many shrubs may be too big for today's small gardens, balconies and window-sills. Begin to train shrubs when they are young. Know how your shrub grows and how its branches should be spaced.

A mixture of nirgudi or chaste tree, lantana, capers, barlerias, spiny asparagus and clerodendrons can give you an impenetrable hedge, or they can be planted to form the background to your garden. This will attract birds to feed and nest, butterflies and moths to breed and feed, and several other creatures to seek shelter. Emigrant butterflies will come to lay eggs on the cassias, and plants like rattlepods, turnsoles and other borages will bring a host of Milkweed butterflies to cluster around them. Specific butterfly food plants like birthworts, passionflowers, curry-leaf and limes will bring butterflies to lay eggs on them. Common sorrel is a good ground cover and Pale Grass Blue butterflies will visit it to lay eggs. Similarly, bankmat is a good ground cover and its flowers attract butterflies. Ixora will not only brighten the garden with its scarlet flowers, but also attract butterflies for nectar. Balsam beds will bring nectar seeking Hawkmoths hovering over the flowers. Screw fruit bush and fire bush will surely attract sunbirds and other nectar seekers.

Natural Pruning

A wild flower garden should always be informal in style, therefore put away those shears. Shrubs are not to be shaped into balls and boxes. Most shrubs trimmed this way lose all their natural beauty. Of course, pruning is necessary for some plants, but on a moderate scale. You can cut back long branches to maintain the size of the shrub. Reach well into the shrub and remove branches to keep the shrub open and airy.

Mulching and Watering

Organic mulch in beds helps retain moisture and prevents other unwanted plants from germinating. Dry hay, wood shavings or shredded bark is a good choice, because it is easy to walk on and doesn't wash away in the rain. If you prefer to use a rock mulch, the best choice is river rock, at least 2-3 cm in diameter. Smaller sized rock can easily work itself into the soil or be washed aside during heavy rain.

In the absence of rain, watering is essential. Perennial shrubs need far less water than annuals. For a large garden, a good sprinkler system or a drip system of flexible black tubing may be ideal.

Manuring

Wild plants do not require heavy manuring, but it is essential to maintain the soil quality by adding adequate organic manure at least once a year. Well rotted farm-yard manure or kitchen compost should be mixed thoroughly in a proportion of 50% to a 15 cm depth into the top soil. Around the same time, bedding for the annuals should be prepared. Excessive manuring will cause more leafy growth and fewer flowers.

These are just broad guidelines to get you started. A wild flower garden is certainly a happy place to be in, as it provides a place of relaxation and recreation for the gardener and the naturalist in you. With very little care, the wild flower garden can soon become your own little wilderness at your very doorstep.

THE COLOURFUL ARRAY

RANUNCULACEAE   **Buttercup Family**
1. **DECCAN CLEMATIS** *Clematis triloba*
   Names: Sans: *Laghuparnika*, Hin: *Murhar*, Mar: *Morvel*, *Ranjaee*, Kan: *Merhari*.
   Flower: 2-3 cm across. **Distr.**: Peninsular and Central India. **Habit**: A spreading climber
   with perennial rootstock, among shrubs and grasses along forest paths, mainly on hills.
   Profuse flowering makes this climber conspicuous in forests. **Flowering**: September -
   December.

NYMPHAEACEAE   **Water lily Family**
2. **STAR WATER LILY** *Nymphaea nouchali* var. *cyanea*
   Names: Hin: *Nilkamal*, Mar: *Krishnakamal*, Guj: *Nilkamal* Beng: *Nilshapla*, Tel:
   Nallakalava, Tam: *Nilotpalam*, Mal: *Sitambel*. **Flower**: 8-15 cm across. **Distr.**: India.
   Habit: Faintly fragrant flowers open throughout the day. Gregarious herb with floating
   leaves. Seen in lakes, ponds and canals. Leaf margin entire or bluntly toothed. Flowers
   pale purple or blue. **Flowering**: January-December.

3. **COMMON WATER LILY** *Nymphaea pubescens*
   Names: Hin: *Kanval*, Mar: *Kamal*, Guj: *Nilo phul*, Beng: *Kanval*, Tam: *Allitamarai*,
   Kan: *Nyadale huvu*, Mal: *Neerambal*, Raj: *Be*. **Flower**: 8-20 cm across. **Distr.**: Warmer
   regions of India. **Habit**: Gregarious in shallow ponds. Large floating leaves velvety,
   prominently veined on under surface. Leaf margin sharply toothed. Flowers open in the
   evening and close next morning. Flowers red, pale rose or white. **Flowering**: January-
   December.

NELUMBONACEAE   **Lotus Family**
4. **INDIAN LOTUS** *Nelumbo nucifera*
   Names: Sans, Beng, Ory: *Padma*, Hin, Mar: *Kamal*, Guj: *Suriyakamal*, Tel: *Kalung*,
   Tam: *Ambal*, Kan: *Kamala*, Kash: *Pamposh*, Mal: *Thamara*. **Flower**: 10-25 cm across.
   Distr.: Throughout India (Himalaya between 600-1,400 m, Kashmir to Uttar Pradesh)
   and Nepal. **Habit**: Large saucer-shaped leaves often emerge out of the water. Flowers
   fragrant, ranging from pink to white in colour. Gregarious and often dominant. **Flowering**:
   January-December.

PAPAVERACEAE   **Poppy Family**
5. **MEXICAN POPPY** *Argemone mexicana*
   Names: Sans, Tel: *Brahmadandi*, Hin: *Bharband*, Guj: *Darudi*, Raj: *Satyanasi*, Mar:
   Pivala dhotara, Beng: *Siyal kanta*, Tam: *Kudiyoetti*, Kan: *Datturi*, Mal: *Ponnummatum*.
   Plant: 80 cm. **Flower**: 2.5-7.5 cm across. **Distr.**: Native of West Indies, naturalised in
   India. **Habit**: Gregarious, prickly invader seen in degraded lands. **Miscellaneous**: Oil
   from seeds is a harmful adulterant in mustard oil. **Flowering**: January-December.

CAPPARACEAE   **Caper Family**
6. **BARE CAPER** *Capparis decidua*
   Names: Sans: *Karira*, Hin: *Kareel*, Mar, Guj: *Ker*, Raj: *Kerro*, Tel: *Kariramu*, Tam:
   Sengam, Kan: *Chippuri*. **Plant**: 6 m. **Flower**: 2.5 cm across. **Distr.**: Drier parts of North
   India, Deccan Peninsula. **Habit**: A straggly, densely branching thorny shrub or small tree
   with a few leaves on young shoots. **Miscellaneous**: Food plant of Pierid butterflies.
   Seeds dispersed by birds. Unripe fruit is pickled. **Flowering**: March-October.

7. **HEDGE CAPER** *Capparis sepiaria*
   Names: Hin: *Heens, Kanthari,* Mar: *Kanthar.* **Plant**: 3-6 m. **Flower**: 0.8-1 cm across. **Distr.**: India (except arid northwest), Sri Lanka. **Habit**: A spreading, wiry, much branched shrub with recurving spines, common along roads and clearings. Locally abundant. **Miscellaneous**: Caterpillars of Pierid butterflies like Gulls, Pioneers, Great Orange Tip and Yellow Orange Tip feed on its leaves. Birds disperse seeds. Planted as hedge plant. Plant used in traditional medicine. **Flowering**: February-May.

8. **CEYLON CAPER** *Capparis zeylanica*
   Names: Sans: *Karambha,* Hin: *Ardanda,* Mar: *Govindi,* Beng: *Kalokera,* Tel: *Adonda,* Tam: *Adondai,* Kan: *Totulla,* Punj: *Hees.* **Plant**: 4-8 m. **Flower**: 4 cm across. **Distr.**: India (except arid Northwest); Himalaya up to 1,000 m, Kashmir to Uttar Pradesh), Nepal, Sri Lanka. **Habit**: A climbing shrub with hooked spines. Common in the forested tracts. Flowers are cream coloured in the morning, then change from pinkish to red and are purple by evening. **Miscellaneous**: Monkeys, civet cats, and squirrels disperse seeds. Caterpillars of Pierid butterflies like Gulls, Pioneers, Great Orange Tip and Yellow Orange Tip feed on this plant. Fruit is pickled. **Flowering**: October - April.

CLEOMACEAE   **Spider flower Family**

9. **WHISKERED SPIDER FLOWER** *Cleome gynandra*
   Names: Hin: *Aajgandha,* Raj: *Safed bagro.* **Plant**: 80 cm. **Flower**: 2.5 cm across. **Distr.**: Throughout India, Sri Lanka. **Habit**: Prefers drier open areas, where this annual herb is seen along village roads and rubbish dumps. A ready coloniser around cultivation. Seeds and leaves are used in traditional medicine. **Flowering**: June-September.

10. **COMMON SPIDER FLOWER** *Cleome rutidosperma*
    Plant: 50-70 cm. **Flower**: 0.8-1.5 cm across. **Distr.**: India (except arid Northwest). **Habit**: A gregarious, prostrate, spreading annual herb, also seen growing erect and tall. This highly adaptable plant has become widespread in recent years. **Flowering**: August-January.

11. **YELLOW SPIDER FLOWER** *Cleome viscosa*
    Names: Sans: *Arkakanta,* Hin: *Hurhur,* Mar: *Kanputi,* Guj: *Talvani,* Raj: *Bagro,* Beng: *Hurhuria,* Tel: *Kukhavoninta,* Tam: *Nayikkadugu,* Kan: *Naibela,* Mal: *Ariavala.* **Plant**: 1m. **Flower**: 1.5 cm across. **Distr.**: India, Nepal. **Habit**: An erect, hairy annual, commonly grows gregariously along roadsides, degraded open land and forest edges. **Miscellaneous**: Leaves used in traditional medicine. **Flowering**: April-October.

PORTULACACEAE   **Purselane Family**

12. **COMMON PURSELANE** *Portulaca oleracea*
    Names: Sans: *Lonika,* Hin: *Khursa, Badi-noni,* Mar: *Bhuigoli,* Guj: *Moti loni,* Raj: *Luni,* Beng: *Baraloniya,* Tel: *Peddapayilikira,* Tam: *Karikeerai,* Kan: *Doddagooni soppu,* Mal: *Kariecheera,* Ory: *Puruni sag.* **Plant**: 15-30 cm. **Flower**: 8 mm across. **Distr.**: India, Nepal. **Habit**: This small, low growing, succulent herb with reddish stems is a common roadside plant and intruder in gardens. Short-lived flowers remain open till mid-day. **Miscellaneous**: Caterpillars of Great and Danaid Eggfly butterflies feed on this plant. Leaves used in salad and stems pickled. Stems and leaves used in traditional medicine. **Flowering**: January-December.

13. **WILD LADIES' FINGERS** *Abelmoschus manihot* ssp. *tetraphyllus*
    Names: Hin: *Jangli bhindi*, Mar: *Ran bhendi*, Guj: *Kantalo bhende*, Assam: *Usipak*.
    Plant: 2.7 m. **Flower**: 5-7.5 cm across. **Distr.**: India (Himalaya up to 2,400 m, Uttar
    Pradesh to Arunachal Pradesh) absent in drier regions, Nepal and Bhutan. **Habit**: A tall
    bristly annual, very common on degraded land and along roads. Closely related to the
    cultivated *bhendi* (*A. esculentus*). **Miscellaneous**: Tussocked caterpillars of the
    Lymantridae moths feed on the plant. The scientific name *Abelmoschus* is derived from
    the Arabic *abu-al-mosk* (father of musk) referring to the musk-scented seeds of the
    cultivated musk mallow (*A. moschatus*). **Flowering**: August-November.

14. **COUNTRY MALLOW** *Abutilon indicum*
    Names: Hin: *Kanghi*, Raj: *Tara-kanchi*, Mar: *Petari*, Beng: *Potari*, Tel: *Tuturabenda*,
    Tam: *Paniyaratutti*, Kan: *Tutti*, Mal: *Velluram*. **Plant**: 1-2 m. **Flower**: 2-3 cm across.
    Distr.: Throughout warmer parts of India. **Habit**: Usually a perennial, this shrub is
    often dominant on disturbed land. Heart-shaped pointed leaves are silky soft. Flowers
    open in the evening. **Miscellaneous**: Caterpillars of Danaid Eggfly butterflies feed on
    this plant. Stem fibres used for making rope. Leaves used in traditional medicine.
    Flowering: September-April.

15. **PERSIAN MALLOW** *Abutilon persicum*
    Names: Hin: *Chota banse, Tepari*, Mar: *Madam*, Mal: *Thutthi*. **Plant**: 2.1 m. **Flower**: 4 cm
    across. **Distr.**: India (absent in drier regions). **Habit**: This erect, hairy shrub prefers to
    grow on forested hills between 400-1,350 m. Commonly seen along the ghat roads,
    where it is a ready coloniser. **Miscellaneous**: Stem yields silky fibre. **Flowering**:
    November-January.

16. **COMMON MALLOW** *Azanza lampas*
    Names: Hin: *Jangli bhindi*, Mar: *Ran-bhendi*, Guj: *Jungli paras piplo*, Tel: *Adavipratti*,
    Kan: *Turuve*, Mal: *Katthurparathi*, Ory: *Bilo kapasiva*, Assam: *Bon kapas*. **Plant**: 0.9-
    2 m. **Flower**: 5-7 cm across. **Distr.**: Well-wooded regions of India (Himalaya up to 1,500 m,
    Uttar Pradesh), W. and C. Nepal. **Habit**: A tall, lanky, roadside shrub with rough leaves.
    Prefers forest edges and hilly terrain. Lower leaves have three lobes, while the flowering
    stem may have leaves with single lobe. **Miscellaneous**: Flowers are eaten by the
    voracious red and black spotted blister beetles (*Mylabris* sp.). **Flowering**: October-
    April.

17. **LESSER WHITE MALLOW** *Hibiscus hirtus*
    Names: Mar: *Dupari*. **Plant**: 1.0 m. **Flower**: 2.5 cm across. **Distr.**: Lower forests of
    Western Ghats and Deccan. **Habit**: This erect hairy mallow has irregularly toothed
    leaves, of variable shapes. Prefers moist deciduous to semi-evergreen forests where it is
    seen on the hills and in forest clearings. Flowers are usually white, occasionally pink.
    Flowering: October-February.

18. **GRAPE-LEAVED MALLOW** *Kosteletzkya vitifolia*
    Names: Mar: *Van kapus*. **Plant**: 1.5 m. **Flower**: 4.5-10 cm across. **Distr.**: Throughout
    the moist well wooded warmer regions of India. **Habit**: Soft hairy, 3-7 lobed leaves of
    this lanky herb are typical of this annual. Common along forest roads. **Miscellaneous**:
    Flowers are visited by nectar seeking insects and petals are relished by the red and black
    spotted blister beetles (*Mylabris* sp.). **Flowering**: September-December.

19. BRAZIL JUTE *Malachra capitata*
Names: Hin: *Vilayati bhindi*, Mar: *Ran bhendi*, Guj: *Pardesi bhindo*, Beng: *Ban bhindi*,
Plant: 1.5 m. **Flower**: 1.8-2 cm across. **Distr.**: Native of Brazil, naturalised in India.
Habit: A coarse, hairy annual introduced from Brazil as a fibre plant, now an aggressive
invader. Grows gregariously near marshy low-lying land. **Miscellaneous**: Flowers attract
butterflies and other insects. Yields fibre resembling jute. **Flowering**: Sept.-Dec.

20. COMMON SIDA *Sida acuta*
Names: Hin: *Bala*, Mar: *Tupkari*, Guj, Raj: *Bala*, Beng: *Pila-barelashikar*, Tel:
Neelabenda, Tam: *Vattatirippi*, Kan: *Cheru paruva*. **Plant**: 1 m. **Flower**: 10 mm across.
Distr.: India. **Habit**: A shrub with slender branches seen commonly along roadsides.
Dominant in degraded land. **Miscellaneous**: Flowers attract butterflies. Fibre used as
substitute jute. Leaves and roots used in traditional medicine. **Flowering**: January-
December.

21. ANGLED SIDA *Sida rhombifolia*
Names: Hin: *Atibala*, Mar: *Sahadevi*, Guj: *Khetraubat-atibala*, Beng: *Lalbarela*, Tel:
Attibalachettu, Tam: *Chitramutti*, Kan: *Binnegarugagida*, Mal: *Anakurunthothi*. **Plant**:
1.5 m. **Flower**: 0.8-2 cm across. **Distr.**: Throughout India, Nepal. **Habit**: A small
gregarious shrub, that prefers forest edges and hillsides. Flowers open by mid-day.
Miscellaneous: Food plant for caterpillars of Lemon Pansy butterfly. Flowers attract
butterflies. Stems and roots are used in traditional medicine. **Flowering**: September-
February.

22. COMMON PURPLE MALLOW *Urena lobata*
Names: Hin: *Bachita*, Mar: *Van-bhendi*, Beng: *Okhra*, Tel: *Peddabenda*, Tam: *Ottatti*,
Kan: *Otte*, Mal: *Uram*. **Plant**: 1.5 m. **Flower**: 1.5 cm across. **Distr.**: Well-wooded regions
of India (Himalaya; Kashmir to Arunachal Pradesh), Pakistan, Nepal and Bhutan. **Habit**:
An erect perennial, with angled, lobed and toothed leaves. Seen along roadsides on the
hills up to 1,800 m. **Miscellaneous**: Seed capsule covered with bristles, which catch on
to animal fur to help in seed dispersal. Cultivated for termite- and water-resistant fibre
used in making fishing line, rope and binding cord. Roots and flowers used in traditional
medicine. **Flowering**: January-December.

STERCULIACEAE · **Sterculia Family**

23. SCREW FRUIT BUSH *Helicteres isora*
Names: Sans: *Mriga shinga*, Hin, Punj, Raj: *Maror phali*, Mar: *Murud sheng*, Guj:
Marad sing, Beng: *Atmora*, Tel: *Nuliti*, Tam: *Valampiri*, Kan: *Yedamuri*, Mal: *Kaivum*,
Ory: *Murmuria*. **Plant**: 3 m. **Flower**: 2.5 cm across. **Distr.**: India, Sri Lanka. **Habit**: A
shrub or small tree, common on forested hill slopes. Spirally twisted fruits are
unmistakable. On hills up to 1,000 m. Flaming red flowers fade to dull bluish-grey.
Miscellaneous: Favourite with nectar feeding birds and butterflies. Caterpillars of
Marumba dyras, a hawkmoth, Common Sailor and Golden Angle Skipper butterflies
feed on the plant. Dried fruit and bark is used in traditional medicine. **Flowering**: July-
September.

24. SCARLET PHOENICIAN *Pentapetes phoenicea*
Names: Mar: *Dupari*. **Plant**: 0.6-1.5 m. **Flower**: 2.5-3.5 cm across. **Distr.**: Native of
Bengal, naturalised in other regions as garden escape. **Habit**: An annual, erect herb, often
seen in backyard gardens and around garbage dumps. Solitary flowers open at noon and
fade next morning. **Flowering**: August-November.

TILIACEAE
Jute Family

25. COMMON BURBUSH *Triumfetta rhomboidea*
Names: Sans: *Jhinjharita*, Hin: *Chiki*, Beng: *Ban okra*, Mar: *Jhinjhira*, Guj: *Jhipato*, Tel: *Kadubende*, Ory: *Bojoramuli, Jolajuti*. Plant: 1-2 m. Flower: 1.2 cm across. Distr.: India, Nepal, Sri Lanka and Pakistan. Habit: This erect annual with drooping branches prefers well-wooded regions from plains to the hills up to 950 m. Lower leaves are three lobed, while those on the flowering stem have single pointed apex. Flowers open by late mid-day. Seen commonly along the roads after monsoon. Miscellaneous: Mammals disperse bristly seed capsules that catch on to their fur. Flowering: August-December.

LINACEAE
Linseed Family

26. MYSORE LINSEED *Linum mysorensis*
Names: Mar: *Undri*. Plant: 15-45 cm. Flower: 8 mm across. Distr.: Peninsular India, Sri Lanka. Habit: A dainty little annual herb seen among short grasses. Linseed is also called flax. Common on open hill slopes just after the rains. Flowering: October-December.

ZYGOPHYLLACEAE
Caltrop Family

27. PUNCTURE PLANT *Tribulus terrestris*
Names: Hin: *Gokhru*, Mar: *Sarrata*, Guj: *Betagokhru*, Raj: *Kanti*, Beng: *Gakhura*, Tel: *Chinnipalleru*, Tam: *Nerunji*, Kan: *Sanna neggilu*, Mal: *Nerunji*. Plant: 1.6 m. Flower: 1.5 cm across. Distr.: India, Sri Lanka. Habit: A low growing branching herb, seen from coastal plains up to 500 m on the hills. Prefers dry, open country. Spiny fruits often puncture bicycle tyres, hence the common English name. Miscellaneous: Tender shoots eaten. Roots and fruit are used in traditional medicine. Flowering: January-December.

GERANIACEAE
Geranium Family

28. AFRICAN GERANIUM *Monsonia senegalensis*
Names: Tel: *Nirkancha*. Plant: 10-20 cm. Flower: 2 cm across. Distr.: India (drier regions of Maharashtra, Andhra Pradesh, NW Rajasthan, Gujarat). Habit: A low growing erect annual herb seen among gravelly embankments along the canals and wells. Flowers are usually solitary. Flowering: August-November.

OXALIDACEAE
Wood-sorrel Family

29. LITTLE TREE PLANT *Biophytum sensitivum*
Names: Sans: *Viparitalajju*, Hin: *Alm bhusha, Lajjalu*, Tam: *Tintanali*, Mal: *Mukkutti*. Plant: 6-10 cm. Flower: 8 mm across. Distr.: India (except drier NW), Nepal. Habit: An annual herb, prefers moist shaded roadsides and backyards. Up to 1,400 m on the hills. Miscellaneous: Plant used in traditional medicine. Flowering: August-January.

30. COMMON SORREL *Oxalis corniculata*
Names: Hin: *Amrit sak*, Mar: *Ambuti*, Raj: *Khatari*, Beng: *Amrit sak*, Tel: *Pulichinta*, Tam: *Puliyarai*, Kan: *Hulichikkai*, Mal: *Puliyarel*, Punj: *Amlika*. Plant: 6-25 cm across. Flower: 9 mm. Distr.: Native of Europe, naturalised in India (Himalaya up to 2,800 m). Habit: Common in moist sheltered gardens and cultivation. This creeping herb intrudes into potted plants and lawns. Miscellaneous: Caterpillars of Pale Grass Blue butterfly feed on the plant. Sour leaves used in chutneys. Flowering: January-December.

31. **BROAD-LEAVED SORREL** *Oxalis latifolia*
Names: Hin: *Khattimeethi*. **Plant**: 20 cm. **Flower**: 1.2 cm across. **Distr.**: Native of Central and tropical South America, naturalised in India, Nepal. **Habit**: A stemless, perennial herb with underground bulb. Leaflets are broadly rounded and triangular. Often overlooked in gardens where it is an intruder. More common on hills, among plantations up to 1,200 m. **Flowering**: June-April.

BALSAMINACEAE   Balsam Family

32. **ROCK BALSAM** *Impatiens acaulis*
Plant: 15 cm. **Flower**: 2.5-3 cm across. **Distr.**: Peninsular India, Sri Lanka. **Habit**: Occurs in small or large clusters in the hills up to 750-2,050 m on dripping wet rock faces. Perennial tuberous rootstock is anchored precariously on vertical rock faces. Flowers are showy, and very pretty among this group. The plant can be mistaken for begonia, but for the distinct slender, long, curved spur of the flower. Locally endangered. **Flowering**: August-November.

33. **COMMON BALSAM** *Impatiens balsamina'*
Names: Sans: *Dushpatrijati*, Hin, Guj: *Gulmehndi*, Mar: *Terda*, Beng: *Dupati*, Tam: *Kasittumbai*, Mal: *Mecchingom*, Punj: *Bantil*. **Plant**: 30-90 cm. **Flower**: 2 cm across. **Distr.**: India, Sri Lanka. **Habit**: This erect succulent annual is common along roadsides, forest roads and sunny hill slopes up to 600-1,250 m. Very gregarious in habit, it is seen throughout the well wooded, moist regions of India. **Miscellaneous**: Caterpillars of White Striped Hawkmoth (*Theretra oldenlandia)*, Green Striped Hawkmoth (*Rhyncholaba acteus)* and hairy caterpillars of Arctiid moths feed on this plant. Flower extract is known to have antibacterial and antifungal properties. **Flowering**: August-October.

34. **DALZELL'S YELLOW BALSAM** *Impatiens dalzellii*
Plant: 25-45 cm. **Flower**: 1.2 cm across. **Distr.**: Western Ghats (Maharashtra). **Habit**: Exclusively found on hills of the Western Ghats above 1,000 m. Though not gregarious, it is often seen in small groups. Easy to recognise, as it is the only yellow flowering balsam in its range. **Miscellaneous**: *Impatiens* in Latin means impatient, referring to the explosive release of the seed when a ripe capsule is touched. **Flowering**: August-October.

35. **WESTERN HILL BALSAM** *Impatiens pulcherrima*
Plant: 45-60 cm. **Flower**: 4-6 cm across **Distr.**: Western Ghats (Maharashtra). **Habit**: Stout succulent herb, mostly annual, seen on the shaded, sheltered slopes of forested hills above 1,000 m. Unlike other balsams, it is not gregarious. It has broad, showy flowers. **Flowering**: August-October.

36. **ROSEMARINE HILL BALSAM** *Impatiens rosmarinifolia* (= *I. oppositifolia*)
Plant: 30-45 cm. **Flower**: 1.3 cm across. **Distr.**: Western Ghats, Sri Lanka. **Habit**: A slender, drooping annual, this gregarious herb is commonly seen on the hills on open grassy slopes up to 1,220 m. Grows gregariously in suitable locations. Purple flowers, sometimes with orange tinge. **Flowering**: August-October.

46

VITACEAE   Vine Family

37. **WOODROW'S GRAPE TREE** *Cissus woodrowii*
    Names: Mar: *Girnul*. **Plant**: 1.5-2 m. **Distr.**: Western Ghats, Deccan (Maharashtra).
    Habit: An erect woody shrub with rough grey fissured bark. Seen mainly on the forested
    hillsides. **Miscellaneous**: Flowers very attractive to butterflies, bees and other insects.
    Generic name is derived from Greek kissos (ivy). **Flowering**: June-July.

FABACEAE (PAPILIONACEAE)   Pea Family

38. **CRAB EYED CREEPER** *Abrus precatorius*
    Names: Sans: *Gunja*, Hin: *Ratti*, Mar: *Gunj*, Guj: *Chanoti*, Raj: *Chirmi*, Beng: *Kunch*,
    Tel: *Guriginja*, Tam: *Gundumani*, Kan: *Guluganji*, Mal: *Kunni*. **Plant**: 5 m. **Flower**: 7-
    9 mm long. **Distr.**: India, Sri Lanka, Pakistan. **Habit**: Usually an annual, but at times
    perennial. This climber is seen on thickets around forest edges from coast to plains up to
    900 m on the hills. Prefers deciduous forests. Flowers are pink to almost white. Bright
    black and red seeds are unmistakable. **Miscellaneous**: Caterpillars of Common Cerulean
    and Indian Sunbeam butterflies feed on the plant. Dried leaves (*hari patti*) eaten with
    paan (betel leaf). The seeds are toxic, and used for ornamental purposes. **Flowering**:
    September-November.

39. **COMMON SWORDBEAN** *Canavalia gladiata*
    Names: Hin: *Khadsampal, Bara sem*, Mar: *Abai*, Beng: *Makhan shim*, Tel: *Year tamma*,
    Tam: *Segapu thambattai*, Kan: *Shembiavare*. **Flower**: 2.5-4 cm across. **Distr.**: Peninsular
    India. **Habit**: A stout perennial or biennial twiner, becomes conspicuous as it flowers.
    Seen on thickets at forest edges. Common in deciduous forests. **Miscellaneous**:
    Caterpillars of Common Sailor butterflies feed on the plant. Flowers favoured by carpenter
    bees. Tender pods are cooked as a vegetable and the seeds are eaten. Pods are used in
    traditional medicine. **Flowering**: August-October.

40. **BOMBAY BEAN** *Clitoria annua* (= *C. biflora*)
    Plant: 35-50 cm. **Flower**: 2.5 cm across. **Distr.**: Endemic to Western Ghats (Maharashtra).
    Habit: An erect, softly hairy herb, usually having two flowers at each node (junction of
    the leaf and stem). Common name is derived from its being endemic to the erstwhile
    Bombay Presidency. **Flowering**: August-October.

41. **BUTTERFLY BEAN** *Clitoria ternatea*
    Names: Hin, Beng: *Aparajita*, Mar: *Gokurna*, Tam: *Kakkanam*. **Plant**: 6 m. **Flower**: 4-
    6 cm long. **Distr.**: Throughout India. **Habit**: A slender vine commonly seen on wayside
    hedges, thickets and scrub forest, from plains to coast up to 750 m on the hills. Often
    cultivated in gardens. Flowers are occasionally white. **Miscellaneous**: Roots and seeds
    are used in traditional medicine. **Flowering**: June-January.

42. **CREEPING HEMP** *Crotalaria filipes*
    Plant: 15-37 cm. **Flower**: 5 mm across. **Distr.**: Western Ghats and Deccan (Maharashtra,
    Karnataka). **Habit**: A low growing, silky hairy, creeping herb, radiating in different
    directions. However, among dense grass it grows erect. Seen in forest clearings and on
    roadsides. **Miscellaneous**: Flowers attract small butterflies and other small nectar-
    seeking insects. **Flowering**: August-December.

43. **GREATER RATTLE POD** *Crotalaria leschenaultii*
Names: Sans: *Ghantarava*, Hin: *Jhunjhunia*, Mar: *Dingla, Dayali*, Beng: *Pipuli jhunjhun.*
Plant: 1-1.5 m. **Flower**: 2 cm across. **Distr.**: India, in suitable habitats. **Habit**: This erect, branching shrub with woody stems prefers forested hilly regions. Conspicuous along the hill roads when flowering. **Miscellaneous**: Flowers favoured by carpenter bees. Bruised and dried plant is very attractive to milkweed butterflies like Tigers and Crows, as a source of an alkaloid essential for their reproduction. Yields a fairly strong fibre. **Flowering**: November-March.

44. **COMMON RATTLE POD** *Crotalaria retusa*
Names: Sans: *Shanar ghantika*, Hin: *Khunkhunia*, Mar: *Ghagri*, Beng: *Bil jhunjhun.* Tel: *Pottigilligichcha*, Tam: *Kilukilluppai.* **Plant**: 60-100 cm. **Flower**: 2-3.2 cm across. **Distr.**: India, Sri Lanka. **Habit**: Large, showy flowers make this annual shrub quite conspicuous. Gregarious patches of this shrub are common along river banks and fields in the plains as well as on the hills up to 1,200 m. **Miscellaneous**: Bruised and dried plant is very attractive to milkweed butterflies like Tigers and Crows, as a source of an alkaloid essential for their reproduction. Flowers favoured by carpenter bees. Leaves used in traditional medicine. **Flowering**: September-April.

45. **COMMON PSORALEA** *Cullen corylifolia* (=*Psoralea corylifolia*)
Names: Hin: *Babchi*, Beng: *Babachi*, Mar, Guj: *Bavchi*, Tel: *Baavanchalu*, Tam: *Kaarboka arisi*, Kan: *Bavanchigida*, Mal: *Karpokkari*, Ory: *Bakuchi.* **Plant**: 0.6-1m. **Flower**: 4 mm across. **Distr.**: India, Sri Lanka. **Habit**: An erect annual shrub, commonly seen around village dumps, cultivation and roadsides. **Miscellaneous**: Seeds and roots are used in traditional medicine. **Flowering**: July-December.

46. **FOREST BEANSTALK** *Derris scandens*
Names: Beng: *Noalata*, Hin: *Gonj*, Mar: *Mota-sirili*, Mal: *Mujal-valli*, Ory: *Kamocho*, Punj: *Gunj*, Tam: *Anaikkathu*, Tel: *Choratali-badu.* **Plant**: 10 m. **Flower**: 6 mm across; flower raceme: 25-46 cm. **Distr.**: Peninsular and Central India to West Bengal, Sri Lanka. **Habit**: Extensively spreading, large, woody climber that can be seen growing on trees up to 30 m in forests. From coast to plains and hills up to 750 m. Often locally abundant. Flowers profusely in dense, drooping clusters, white or pale pink. **Miscellaneous**: Used as fish poison. Attracts bees and other insects. **Flowering**: June-August.

47. **MANGROVE BEANSTALK** *Derris trifoliata* (= *D. uliginosa*)
Names: Beng: *Panlata*, Mar: *Karanjvel*, Ory: *Kelia*, Tel: *Nalla tige.* **Plant**: 6 m. **Flower**: 1 cm across; flower spike: 12 cm. **Distr.**: Coastal India (including Andaman and Nicobar Is.), Sri Lanka, Pakistan, Bangladesh. **Habit**: A mangrove associate, it is seen as a straggler or a climbing liana above the high-tide mark. At places, it is locally abundant. **Miscellaneous**: Bark is used as fish poison. Some insecticidal property present. Profuse flowering attracts bees and black ants for nectar. **Flowering**: January-March.

48. **SILKY INDIGO** *Indigofera astragalina* (= *I. hirsuta*)
Plant: 0.6-1.20 m. **Flower**: 5 mm across; flower spike: 5-20 cm long. **Distr.**: India, Sri Lanka. **Habit**: This hardy annual can thrive in poor soil. Usually has 7-9 leaflets. Stem and branches densely clothed with fine long, silky hairs. Seen throughout the plains and on the hills up to 600 m. **Miscellaneous**: Genus name from indigo and Latin *fero* (to bear). Source of indigo dye in West African countries. Leaves used in traditional medicine. **Flowering**: September-February.

49. COMMON COWITCH *Mucuna pruriens*
Names: Hin: *Krainch, Khajkhujli, Kiwach*, Mar: *Khaj-kuili, Kavach*, Guj: *Kivanch,*
Beng: *Alkushi*, Tel: *Dulagondi*, Tam: *Poonaipidukkam*, Kan: *Nasukunni*, Mal: *Naicorna*,
Ory: *Kaincho*. **Flower**: 4 cm across. **Distr.**: India (up to Himalayan foothills), Sri Lanka.
Habit: Annual vine that spreads extensively. Commonly seen on fences, roadsides and
in forests. S-shaped pods are covered with dense velvet-like bristles that cause severe
itch if touched. **Miscellaneous**: Caterpillars of Common Sailor butterfly, Death's Head
Hawkmoth (*Acherontia lachesis*) and *Clanis phalaris*, a hawk moth, feed on this plant.
Seeds and roots are used in traditional medicine. **Flowering**: October-November.

50. INDIAN KUDZU *Pueraria tuberosa*
Names: Hin: *Sural*, Guj: *Vidarikand*, Mar: *Bharda, Ghorbel*, Beng: *Shimia*, Kan: *Dari,
Gumadigida*, Mal: *Mutukku*, Tel: *Darigummadi*, Punj: *Siali*. **Flower**: 1.5 cm across;
flower spike: 25 cm. **Distr.**: India (Himalaya: Uttar Pradesh to Kashmir 300-2,000 m),
C. Nepal, Pakistan. **Habit**: A woody climber with three large unequal sided leaflets.
Flower spikes appear when the climber is leafless. Occurs in forested regions, prefers
hills up to 2,000 m. Flat pod has soft brown bristles. Tubers are very large, weighing up
to 35 kg. **Miscellaneous**: Caterpillars of *Clanis phalaris*, a hawkmoth, feed on this
plant. Large tubers are used in traditional medicine. **Flowering**: March-April.

51. SENSITIVE SMITHIA *Smithia sensitiva*
Names: Hin, Punj: *Odabrini*, Mar: *Kaola*, Beng: *Nullakashina*, Mundari: *Masuri sing.*
Plant: 30-90 cm. **Flower**: 1.3 cm long. **Distr.**: India (except arid regions). **Habit**: A low
growing annual herb, common along the roads as the monsoon trails off. Grows along
cultivation up to 900 m on the hills. Leaves slightly sensitive to touch. **Miscellaneous**:
Flowers are eaten by the voracious red and black spotted blister beetles *Mylabris* sp.
Leaves and shoots are cooked and eaten. Plant is used in traditional medicine. **Flowering**:
August-October.

52. CARIBBEAN STYLO *Stylosanthes hamatus*
Plant: 30 cm. **Flower**: 3 mm long. **Distr.**: Drier regions of Maharashtra, Gujarat. **Habit**:
A low growing, perennial undershrub, common on gravelly soil in open situations.
Flattened pods are hooked at the apex, hence the common name. **Flowering**: August-
January.

53. COMMON TEPHROSIA *Tephrosia purpurea*
Names: Sans: *Sharapunkha*, Hin: *Sarphanka, Dhamasia*, Mar: *Unhali, Sarpankha*,
Guj: *Ghodakan*, Raj: *Biyani*, Beng: *Ban-nil-gachh*, Tel: *Vempali*, Tam: *Kolingi,* Kan:
Empali, Mal: *Kozhenjil*, Ory: *Pokha*, Punj: *Bansa-bansu*. **Plant**: 0.3-1 m. **Flower**:
0.8-1 cm across. **Distr.**: Throughout India. **Habit**: A much branched perennial herb,
usually seen at forest edges and on roadsides. Leaflets 7-13 present. **Miscellaneous**:
Genus name derived from Greek *tephros* (ash-coloured) referring to the colour of the
leaves of some species in this group. Plant and its seed-oil used in traditional medicine.
Flowering: July-January.

54. PURPLE FEATHER BUSH *Uraria rufescens*
Plant: 0.4 - 1.5 m. **Flower**: 6 mm long. **Distr.**: India (up to Himalayan foothills, absent
in arid regions), Sri Lanka. **Habit**: A stout, straggly shrub seen in small clumps in
deciduous forests. Seen on the hills up to 750 m. Bristly folded up pods with 4 or 5
joints are typical. **Flowering**: September-November.

55. WILD MOONG *Vigna radiata*
Names: Hin: *Jungli moong*, Mar: *Jangli mug, Mukani*. **Plant:** 30 cm. **Flower:** 1.5 cm across. **Distr.:** Peninsular and central India. **Habit:** A twining, trailing annual herb commonly seen along forest edges and on the hills up to 1,000 m. Closely related to the cultivated *moong*. **Miscellaneous:** Bees pollinate flowers. Seeds are eaten by local people in times of scarcity. **Flowering:** August-October.

56. INDIAN SWEET PEA *Vigna vexillata* var. *angustifolia* (= *V. capensis*)
Names: Mar: *Halunda*, Khasi (NE): *Jermei-soh-lang-tor*. **Flower:** 2.5 cm across. **Distr.:** Western Ghats, central India, Himalayan foothills, Sri Lanka. **Habit:** Though not fragrant, it is called sweet pea. A perennial twiner, common on the plains and hills up to 1,200 m. **Miscellaneous:** Roots and seeds are eaten by local people. **Flowering:** August-November.

CAESALPINIACEAE   Peacock flower Family

57. CAMEL'S FOOT CLIMBER *Bauhinia vahlii* above Gangtok 20.4-10
Names: Hin: *Maljan*, Mar: *Chambul*, Beng: *Sihar*, Tel: *Adda*. **Flower:** 3.5-5 cm across. **Distr.:** Throughout India in forested hilly regions (Himalaya up to 1,500 m, Kashmir to Sikkim), Bhutan and Nepal. **Habit:** A large, spreading, woody climber seen covering trees extensively. Flowers gregariously during the summer. Leaves resemble a camel's footprint in shape, hence the common name. **Miscellaneous:** Being an aggressive climber, it covers trees to the extent of hindering their growth, hence is often eradicated. Edible seeds used in traditional medicine. Locals use leaves as plates. Bark used to make ropes. Stem produces tanning material. **Flowering:** April-June.

58. CANDLE CASSIA *Cassia alata*
Names: Mar: *Shimai-agase*. **Plant:** 2-5 m. **Flower:** 2.5 cm across. **Distr.:** Native of S. America, now naturalised in India. **Habit:** A robust shrub, planted in gardens and backyards for decorative and medicinal purposes, seen frequently near roadsides and along watercourses. Flowers lack nectar. **Miscellaneous:** Flowers are pollinated by Carpenter bees (*Xylocopa* sp.) that visit for pollen. Caterpillars of Emigrant butterflies (*Catopsila* sp.) feed on this plant. Leaves used in traditional medicine. **Flowering:** October-June.

59. TANNER'S CASSIA *Cassia auriculata*
Names: Hin: *Tarvar*, Mar: *Tarwad*, Guj: *Aval*, Raj: *Anwal*, Tel, Kan: *Tangedu*, Tam: *Avaram*, Mal: *Avara*. **Plant:** 2 m. **Flower:** 3.5 cm across. **Distr.:** India, Sri Lanka. **Habit:** A much branching, spreading shrub that prefers drier regions along roadsides and can grow in the poorest soils. The stipules are unmistakable ear-like (auricle) extensions at the base of each leafstalk near the stem. **Miscellaneous:** Caterpillars of Emigrant butterflies (*Catopsila* sp.) feed on this plant. Earlier it was used as a source of tannin. Cattle do not eat it. **Flowering:** January-December.

60. FEATHER-LEAVED CASSIA *Cassia mimosoides*
Plant: 30-90 cm. **Flower:** 6 mm across. **Distr.:** India (absent in arid regions), Sri Lanka. **Habit:** A slender herb, seen on the forest floor of scrub and degraded forest, less common in the plains, prefers hills up to 1,600 m. **Flowering:** October-February.

61. **SENNA SOPHERA** *Cassia sophera*
Names: Hin: *Kasaunda*, Mar: *Ran takla*, Beng: *Kal ka shunda*, Tel: *Kond ka shinda*, Tam: *Sulari*, Mal: *Pounantakara*. **Plant**: 1-3 m. **Flower**: 2.5 cm across. **Distr.**: Throughout India, Sri Lanka. **Habit**: Annual or perennial shrub along roads and near habitations, often small groups. Now rapidly spreading on degraded land. Each compound leaf has 6-10 pairs of leaflets. **Miscellaneous**: Preferred food plant of Emigrant butterflies. Bark, leaves and seeds used in traditional medicine. **Flowering**: January-December.

62. **POT CASSIA** *Cassia tora*
Names: Sans: *Dadamari*, Hin: *Chakunda*, Mar: *Takla*, Guj: *Kovariya*, Beng: *Chakundi*, Tel: *Tantemu*, Tam: *Tagrai*. **Plant**: 30-90 cm. **Flower**: 1.5 cm across. **Distr.**: India (Himalaya up to 1,500 m, absent in arid Northwest), Nepal, Sri Lanka. **Habit**: Gregarious annual herb seen along roadsides and wasteland. **Miscellaneous**: Caterpillars of Emigrants (*Catopsila* sp.) and Grass Yellow butterflies (*Eurema* sp.) feed on the plant. Tender shoots and leaves are cooked and eaten. **Flowering**: September-December.

63. **FLAMING SPIKE CLIMBER** *Moullava spicata* (=*Wagatea spicata*)
Names: Mar: *Wagati*, *Wakeri*, Tam: *Pulinakagondai*, Kan: *Hoogliganje*. **Flower**: Flower spike: 60 cm. **Distr.**: Western Ghats (Maharashtra, Karnataka). **Habit**: A woody climber armed with backward curving spines, on trees and large shrubs among the forested tracts of the Western Ghats. Endemic to the region. **Miscellaneous**: Caterpillars of Black Rajah, Blue Nawab, Chestnut-Streaked Sailor, Opaque Six-line Blue, Pointed Ciliate Blue butterflies feed on this plant. **Flowering**: October-May.

MIMOSACEAE   **Touch-me-not Family**

64. **PAINTED THORN BUSH** *Dichrostachys cinerea*
Names: Hin, Raj: *Kolai*, Mar: *Sigam kati*, *Yelati*. **Plant**: 4 m. **Flower**: 2 mm across, flower spike: 2-5 cm long. **Distr.**: India, Sri Lanka. **Habit**: A much branched, thorny shrub or small tree, with spreading crown and branchlets having spines at the end. Prefers open, dry country. Occurs from coast to plains up to 300 m on the hills. Grows even in poor soil. Conspicuous showy bi-coloured drooping flower-spikes. **Flowering**: April-September.

65. **TOUCH-ME-NOT** *Mimosa pudica*
Names: Hin: *Chhui-mui*, *Lajvanti*, Mar: *Lajalu*, Guj: *Lajamani*, Beng: *Lajjabati*, Tel: *Attapatti*, Tam: *Tottalvade*, Kan: *Lajja*, Mal: *Thottavadi*, Ory: *Lajkuri*, Assam: *Nilajban*. **Plant**: 45-90 cm. **Flower**: 1 cm across. **Distr.**: India (Himalaya up to 1,900 m, absent in arid Northwest). Probably introduced from Tropical America. **Habit**: A low growing prickly shrub with very sensitive compound leaves, which close together and droop down when touched. Flat, pale brown bristly pods are unmistakable. Genus name derived from Greek *mimos* (a mimic) and *pudica* means shy, referring to the sensitive leaves. **Miscellaneous**: Roots and leaves are used in traditional medicine. **Flowering**: September-October.

66. **SENSITIVE NEPTUNIA** *Neptunia prostrata* (=*N. oleracea*)
Names: Hin: *Chhui-mui*, Mar: *Panilajak*. **Flower**: 1.25 cm across; flower spike: 7.5-15 cm tall. **Distr.**: In freshwater wetlands throughout India, Sri Lanka. **Habit**: A floating aquatic herb having stems covered with soft, swollen spongy floats. Roots profusely from leaf and flowering nodes. Covers the water surface along with other aquatic plants. Leaves are sensitive to touch. **Flowering**: August-March.

COMBRETACEAE   **Terminalia Family**

67. PAPER FLOWER CLIMBER *Calycopteris floribunda*   ?Sikkim April '10
Names: Sans: *Shvetadhataki*, Hin: *Kokoray*, Mar: *Ukshi*, Tel: *Adivijama*, Tam: *Minnargodi*, Kan: *Marasadaboli*, Ory: *Dhonoti*. **Plant**: 0.72-12 m. **Flower**: 3 cm across. **Distr.**: Central, southern and northeast India. **Habit**: A semi-evergreen straggler liana, often seen in gregarious clumps at forest edges and slopes up to 500 m on the hills. Clusters of pale green flowers which turn pale orange and have a papery appearance. Flowers appear in summer. **Flowering**: March-May.

MELASTOMATACEAE   **Blackmouth Family**

68. MALABAR BLACKMOUTH *Melastoma malabathricum*
Names: Mar: *Palore*, Beng: *Gongai*, Tel: *Pattuda*, Tam: *Nakkukaruppan*, Kan: *Ankerki*, Mal: *Kalampatti*. **Plant**: 3 m. **Flower**: 4.5 cm across. **Distr.**: Forested regions of India (except arid NW). **Habit**: Hills and forests along streams. Flowers have alternate yellow and reddish-purple anthers. Flowers may be pink-magenta, purple or white. Fruit stains black, hence the generic name meaning "blackmouth" in Latin. **Miscellaneous**: Food plant of Atlas moth and Grey Count butterfly. **Flowering**: October-March.

LYTHRACEAE   **Henna Family**

69. FIRE BUSH *Woodfordia fruticosa*
Names: Hin: *Dhai*, Mar: *Dhaiti*. **Plant**: 7 m. **Flower**: 9-13 mm long. **Distr.**: India (Himalaya up to 1,800 m, absent in arid NW), Sri Lanka. **Habit**: A much branched shrub, prefers warmer habitats. Common in central and north India. **Miscellaneous**: Tubular flowers attract birds. Dried flowers used for dyeing. **Flowering**: February-June.

TRAPACEAE   **Water chestnut Family**

70. WATER CHESTNUT *Trapa natans* var. *bispinosa*
Names: Hin, Guj, Raj, Tam, Kan: *Singhara*, Mar: *Shingada*, Beng: *Paniphal*, Tel: *Kubyakamu*, Mal: *Karimpolan*. **Flower**: 8 mm across. **Distr.**: India, Nepal, Sri Lanka. **Habit**: An annual, aquatic floating herb seen in still waters. Cultivated for its edible fruit. Flowers tilt sideways after being pollinated. Fruits have 2 spines. **Flowering**: July-November.

PASSIFLORACEAE   **Passion flower Family**

71. COMMON PASSION FLOWER *Passiflora foetida*
Names: Mar: *Ghani-vel*, Tel: *Tellajumiki*, Tam: *Mupparisavalli*, Kan: *Kukkiballi*, Mal: *Chadayan*. **Flower**: 4 cm across. **Distr.**: Native of tropical S. America, naturalised in India. **Habit**: A climbing vine, seen on forest edges and on village outskirts. Leaves when crushed emit a foul smell, hence the Latin name. A ready coloniser, seen on the hills up to 1,200 m. **Miscellaneous**: Food plant of the Tawny Coster butterfly. **Flowering**: November-May.

CUCURBITACEAE   **Cucumber Family**

72. IVY GOURD *Coccinia grandis*
Names: Sans: *Bimba*, Hin: *Bimb, Kachri*, Mar: *Tendli*, Guj: *Ghobe*, Raj: *Gol*, Beng: *Telakucha*, Tel: *Dandakaya*, Tam: *Kovaikai*, Kan: *Tondekai*. **Flower**: 1.5-2.5 cm across. **Distr.**: India, Sri Lanka. **Habit**: Extensively spreading vine, older stems rough, fissured and swollen at intervals. Male and female flowers on separate plants. Common on fences and thickets. Ripe red fruit conspicuous. **Miscellaneous**: Seeds dispersed by birds. Roots, fruit, leaves used in traditional medicine. Cultivated extensively, tender fruit sold as vegetable, used as a meat tenderiser. **Flowering**: January-December.

73. **WILD MUSK MELON** *Cucumis melo* var. *agrestis*
Names: Mar: *Shinde*, Guj, Raj: *Kachari*. **Plant size**: 1-2 m. **Flower**: 0.8-2 cm across.
Distr.: India, Sri Lanka. **Habit**: This coarse haired ground runner is seen throughout
open tracts from plains to hills up to 1,000 m. Male and female flowers are borne on the
same plant. Bitter, light green fruit with green patches in rows. Closely related to
cultivated variety of musk melons. **Miscellaneous**: Fruit pulp used in traditional medicine.
Flowering: July-November.

74. **SPINY MELON** *Cucumis prophetarum*
Names: Hin: *Khar-indrayan*, Mar: *Kante-indrayan*, Guj: *Kantalan-indranan*, Raj: *Khat-kachario*. **Flower**: 3-4 mm long. **Distr.**: India (drier Northwest Rajasthan, Gujarat,
Maharashtra). **Habit**: Seen throughout the drier open regions on the ground or shrubs as
well as on hedges, this climber has separate male and female flowers on the same plant.
Dark green and white striped fruits have soft fleshy spines. **Miscellaneous**: The plant
has emetic and purgative properties. Fruit pulp is toxic to animals. **Flowering**: August-
January.

75. **BRISTLE GOURD** *Momordica dioica*
Names: Hin: *Golkandra, Kaksa*, Mar: *Kantoli*, Guj: *Kantola*, Raj: *Kankero*, Beng: *Ban
karela*, Tel: *Agakara*, Tam: *Tholoopavai*, Kan: *Karlikai*, Punj: *Kakaura*, Assam: *Bhat-
karela*. **Flower**: 2-3.5 cm across. **Distr.**: India, Sri Lanka. **Habit**: A perennial climber
found growing in the hilly forested tracts and in the Himalaya up to 1,500 m. Fruit has
soft fleshy bristles. **Miscellaneous**: Collected and sold by tribals for vegetables.
Flowering: June-August.

76. **MADRAS PEA PUMPKIN** *Mukia maderaspatana*
Names: Hin: *Agumaki*, Mar: *Chirati, Pangori*, Guj: *Tindori*, Raj: *Ankh phutani*, Beng:
Bilari, Tel: *Noogudosa*, Tam: *Musumusukkai*, Kan: *Mukkalpeeram*, Punj: *Gwala kakri*.
Flower: 6 mm across. **Distr.**: India. **Habit**: A prostrate or climbing annual herb, commonly
seen on thickets along the roads. Found up to 1,800 m on the hills. Male and female
flowers are on the same plant. Leaves vary in shape and size on the same plant. Ripe red
pea-sized fruits are unmistakable. **Miscellaneous**: Plant used in traditional medicine.
Flowering: August-October.

77. **COMMON FRINGED-FLOWER VINE** *Trichosanthes cucumerina*
Names: Mar: *Jangli-padvel*. **Plant size**: 4-6 m long **Flower**: 1.5 cm across. **Distr.**:
India, Sri Lanka. **Habit**: Common on hedges, trees and shrubs, this annual vine has male
and female flowers on the same plant. Female flowers are solitary. Conical fruit, tapering
at both ends with pointed end, are green with white stripes, and later turn orange to red
when ripe. Flowering: July-October.

78. **GREAT FRINGED-FLOWER VINE** *Trichosanthes tricuspidata*
Names: Hin: *Lal Indrayan*, Mar: *Kaundal, Mukal*. **Flower**: 4-8 cm across. **Distr.**: India
(except arid Northwest), Nepal, Bhutan. **Habit**: Large perennial climber, seen on trees
and shrubs. Prefers well wooded moist places. Occurs on the hills up to 2,500 m. Male
and female flowers are on separate plants. **Miscellaneous**: Seeds are dispersed by
birds. Root is used in veterinary medicine. **Flowering**: May-August.

BEGONIACEAE   **Begonia Family**

79. COMMON BEGONIA *Begonia crenata*
Names: Mar: *Berki, Motiyen*. **Plant**: 25 cm. **Flower**: 1 cm across. **Distr.**: Western Ghats (Maharashtra). **Habit**: An annual, succulent herb, quite common among forested hilly regions, preferring to grow on sheltered moss covered rocks in gregarious groups. Male and female flowers are different. **Miscellaneous**: Plant juice used in traditional medicine. **Flowering**: August-September.

AIZOACEAE   **Shore Purselane Family**

80. SAND AND STONE PLANT *Corbichonia decumbens*
Names: Hin, Raj: *Patthar-chatta*. **Plant**: 70 cm. **Flower**: 7 mm across. **Distr.**: Drier regions of India (Maharashtra, Karnataka, Punjab, Rajasthan and Upper Gangetic Plains). **Habit**: A trailing annual or shortlived perennial herb of the drier plains seen among rocks and sandy soil. Flowers open in the afternoon. **Flowering**: August-March.

81. SHORE PURSELANE *Sesuvium portulacastrum*
Names: Beng: *Jadu palang*, Raj: *Lunio*, Tel: *Vangarreddikura*, Tam: *Vankuru valai*, Mar: *Dhapa*. **Flower**: 8 mm across. **Distr.**: Indian coastline and Sri Lanka. **Habit**: A succulent herb with reddish branches rooting at nodes, seen as a dense mass along the coastline. Flowers open early and fade by noon. Resembles the common purselane. **Miscellaneous**: Used as vegetable and planted as sand binder. **Flowering**: Nov.-March.

APIACEAE   **Coriander Family**
82. KONKAN PINDA *Pinda concanensis*
Names: Mar: *Panda, Pinda*. **Plant size**: 0.5-1m. **Flower**: 2.5-3 cm across. **Distr.**: Western Ghats (Maharashtra). **Habit**: Small groups of this endemic herb are conspicuous on open, grassy hill slopes. Among the monsoon plants, this is one of the early flowering herbs. Fruits are less flat than in most of the other species of pinda. **Miscellaneous**: Bees, ants and other small insects pollinate the flowers. **Flowering**: July-August.

RUBIACEAE   **Coffee Family**
83. FALSE GUAVA *Catunaregam spinarum* (= *Randia dumetorum*)
Names: Hin, Beng: *Mainphal*, Mar: *Ghela*, Guj: *Mindhal*, Tel: *Manga*, Tam: *Marukkallavikay*, Kan: *Kare*, Mal: *Kara*, Ory: *Patova*. **Plant**: 5 m. **Flower**: 2 cm across. **Distr.**: India (Himalaya, Himachal Pradesh to Sikkim, absent in arid Northwest), Nepal, Bhutan, Sri Lanka. **Habit**: Common shrub in forested regions and on the hills up to 1,200 m. Solitary fragrant white flowers, later change to pale yellow. **Miscellaneous**: Hard guava-like fruit are utilised by Guava-Blue butterfly caterpillars to feed and pupate within. Mashed root and unripe fruit are used to poison fish. Flowers attract honeybees. **Flowering**: March-June.

84. GUM GARDENIA *Gardenia gummifera*
Names: Sans: *Pindava*, Hin, Mar, Beng: *Dikemali*, Guj: *Kamarri*, Tel: *Manchibikki*, Tam: *Kambilippicin*, Kan: *Cittubikke*, Ory: *Gurudu*. **Plant**: 2-6 m. **Flower**: 4.5 cm across. **Distr.**: Peninsular India, up to Madhya Pradesh. **Habit**: A shrub or small tree in deciduous forests and on degraded hill slopes up to 1,000 m. White flowers turn yellow before fading, are not fragrant. Leaves have 12 to 18 pairs of lateral veins. **Miscellaneous**: Food plant of Bee Hawkmoth (*Cephondes hylas*). Gummy exudate from tips of shoots and buds is used in traditional medicine. **Flowering**: January-May.

85. JUNGLE FLAME *Ixora coccinea*
Names: Hin: *Rukmini*, Mar: *Pendgul*, Beng: *Rangan*, Tel: *Koranam*, Tam: *Chetti*, Kan: *Kepala*, Mal: *Thethi*. **Plant**: 1 m. **Flower**: 2 cm across. **Distr.**: Peninsular India, Sri Lanka. **Habit**: Branching shrub seen along the moist forests near sea coast. Flowers sometimes yellow or pink. **Miscellaneous**: Food plant of Bee hawkmoth (*Cephondes hylas*). Roots and flowers used in traditional medicine. **Flowering**: January-December.

86. DHOBI'S KERCHIEF *Mussaenda frondosa*
Names: Hin: *Bedina*, Mar: *Bhatkes*, Beng: *Nagballi*, Tam: *Vellaiyalai*, Kan: *Billoothi*, Mal: *Parthole*. **Plant**: 2-6 m. **Flower**: 3.5 cm long. **Distr.**: India (absent in arid Northwest), Sri Lanka. **Habit**: This shrub flowers during mid-monsoon and early winter; but the white bracts may be seen on the plant practically throughout the year. Prefers forested hilly tracts. **Miscellaneous**: Food plant of Commander butterfly. Leaves, fruit are used in traditional medicine. Flowers are eaten as pot herb. **Flowering**: July-November.

87. INDIAN PAVETTA *Pavetta crassicaulis*
Names: Hin: *Kankra*, Mar: *Papdil*, Guj: *Papat*, Beng: *Jui*, Tel: *Duyi papata*, Tam: *Pavatta*, Kan: *Pavati*, Mal: *Pavatta*, Ory: *Phingi*. **Plant**: 3 m. **Flower**: 1.5 cm across. **Distr.**: India (including Andaman, Himalaya, except arid NW), Nepal, Bhutan, Sri Lanka. **Habit**: Prefers forested hilly regions, where it is common on the shady slopes. Seen on the hills up to 1,500 m. **Miscellaneous**: Fragrant flowers attract butterflies and other insects. Seeds are dispersed by birds. Food plant of the Bee Hawkmoth. **Flowering**: March-July.

COMPOSITAE   **Sunflower Family**

88. GOAT WEED *Ageratum conyzoides*
Names: Hin: *Visadodi*, Mar: *Ghanera osadi*, Guj: *Asagandha*, Beng: *Uchunti*, Sind: *Hulan tala*, Tam: *Pampillu*, Kan: *Nayitulasi*, Ory: *Puksunga*, **Plant**: 90 cm. **Flower**: 5 mm across. **Distr.**: Native of S. America, naturalised in India (Himalaya; up to 1,500 m), Nepal, Sri Lanka. **Habit**: Annual intruder in gardens and orchards, common name from its goat-like odour. Flower-heads may be pale blue. **Flowering**: January-December.

89. MEXICAN FLOSS FLOWER *Ageratum houstonianum*
Plant: 60 cm. **Flower**: 1.2 cm across. **Distr.**: Native of Peru, Mexico, naturalised in India, mainly on hills. **Habit**: An erect gregarious herb, seen along stream banks on the hills up to 1,000 m. The fluffy, soft flower heads are borne as dense clusters of many flowers. This is a garden escape. There are several ornamental varieties with flowers in blue, lavender, pink and white. **Flowering**: August-December.

90. COMMON FLOSS FLOWER *Chromolaena odorata* (= *Eupatorium odoratum*)
Names: Hin: *Tivra gandha*. **Plant**: 3 m. **Flower**: 1 cm long. **Distr.**: Native of tropical S. America, naturalised in India, Sri Lanka. **Habit**: An aggressive invader, now rampant due to felling of evergreen forests. Very common along roads and on the hills up to 1,000 m. **Miscellaneous**: Flowers attract butterflies. **Flowering**: December-March.

85

86

87

88

89

90

91. PURPLE BANE *Cyathocline purpurea* (= *C. lyrata)*
Plant: 1.25 m. **Flower**: 4 mm across. **Distr.**: India (including Himalayan region, absent in arid regions), Nepal. **Habit**: This slender, erect, strongly aromatic herb is gregarious in habit. Divided comb-like leaves are unmistakable. On the hills often dominant along the roads, stream beds and rice fields. **Flowering**: January-December, except monsoon.

92. MARSH DAISY *Eclipta prostrata* (= *E. alba*)
Names: Sans: *Kesaraja*, Hin: *Babri*, Mar: *Bhringuraja*, Guj: *Bhangra*, Beng: *Kesuti*, Tel: *Galagara*, Tam: *Garuga*, Kan: *Garagadasoppu*, Mal: *Kyonni*. **Plant**: 50-75 cm. **Flower**: 0.5-1 cm across. **Distr.**: India, Nepal, Sri Lanka. **Habit**: A perennial with an affinity for moist open places. It may be erect or prostrate, with branches rooting at nodes. **Miscellaneous**: Flowers attract insects. Whole plant is used in traditional medicine. **Flowering**: October-May.

93. MADRAS CARPET *Grangea maderaspatana*
Names: Hin: *Mustaru, Bhediachim*, Mar: *Mashipatri*, Guj: *Jhinkimudi*, Beng: *Namuti*, Tel: *Mastaru*, Tam: *Masipathri*, Kan: *Davana*, Mal: *Nilampala*. **Plant**: 10-45 cm. **Flower**: 0.5-1 cm across. **Distr.**: India, Nepal, Sri Lanka. **Habit**: An aromatic annual, commonly seen in flat bunches in harvested fields, dry river and pond beds. **Miscellaneous**: Leaves are used in traditional medicine. **Flowering**: December-May.

94. HILL GYNURA *Gynura cusimbua*
Plant: 1-2 m. **Flower**: 2.5 cm long. **Distr.**: India (Himalaya; Uttar Pradesh to Arunachal Pradesh, Western Ghats and hills in Deccan), Nepal, Bhutan. **Habit**: A tall robust, herb. Locally abundant in moist localities, but not common. Prefers the grassy slopes of hills up to 2,400 m. **Miscellaneous**: Generic name derived from Greek *gyne* (female*)* and *oura* (a tail) referring to the long stigma. **Flowering**: August-November.

95. AMERICAN SOFTHEAD *Lagascea mollis*
Plant: 1 m. **Flower**: 1.5 cm across. **Distr.**: Native of Tropical America, now naturalised in India. **Habit**: An erect annual herb, commonly seen among other monsoon vegetation as the monsoon trails off. More common in open drier regions, where it is spreading rapidly. **Flowering**: August-March.

96. COUNTRY DANDELION *Launaea procumbens*
Names: Hin: *Van-gobi*. **Plant**: 15-60 cm. **Flower**: 1.2-2 cm across. **Distr.**: India, more or less throughout Pakistan. **Habit**: A perennial with flat leaves in rosettes, seen among grassy, open slopes and along roads. **Miscellaneous**: Leaves are eaten in curries mixed with other vegetables. **Flowering**: October-March.

97. CONGRESS GRASS *Parthenium hysterophorus*
Names: Hin: *Gajar ghas*, Mar: *Gajar gavat*. **Plant**: 1 m. **Flower**: 4-7 mm across.
Distr.: Native of West Indies, Central and North America, naturalised in India. **Habit**:
A very aggressive invader that grows prolifically and flowers throughout the year.
Being highly adaptable, it can flourish where nothing else will grow. **Miscellaneous**: A
known allergen of humans and cattle, causes asthma, eczema and contact dermatitis.
Plant is used in traditional medicine. **Flowering**: January-December.

98. PURPLE HEADS *Phyllocephalum scabridum* (= *Centratherum phyllolaenum*)
Plant: 60 cm. **Flower**: 1.5-2 cm across. **Distr.**: Western Ghats (Karnataka, Maharashtra,
Gujarat) up to Mt. Abu (Rajasthan). **Habit**: An erect branching annual herb seen
together with other monsoon annuals along the roads on sunny slopes. Occurs mainly
on the hills. **Flowering**: October-November.

99. GRAHAM'S GROUNDSEL *Senecio grahami*
Names: Mar: *Sonki*. **Plant size**: 0.7-1 m. **Flower**: 1.5-2.7 cm across. **Distr.**: Western
Ghats, Maharashtra. **Habit**: This extremely gregarious annual prefers sunny hill slopes,
which become conspicuous during mass flowering. Often seen on old roofs and in the
forks of trees where debris has accumulated. Endemic to the Western Ghats. This herb
is named after John Graham (1805-1839), Deputy Postmaster General, who made a
significant contribution to the flora of Western India, before he died at the age of 34.
Generic name derived from Latin *senex* (an old man), referring to the fluffy, white seed
heads of several species in this group, except this particular species. **Flowering**: August-
December.

100. COMMON SOW THISTLE *Sonchus oleraceus*
Names: Hin: *Dudhi*, Mar: *Pathari*, Tel: *Ratrinta*. **Plant**: 0.3-1 m. **Flower**: 1.8 cm
across. **Distr.**: India (including Himalaya), Nepal. **Habit**: Erect annual herb with grooved
stem and long, toothed leaves deeply cut almost to the midrib. Narrow ear-like extensions
of the leaves clasp around the stem. Each flower has a tall stalk. Pure white silky seed
heads seen after flowering. Usually seen in temperate and sub-tropical regions. On the
hills up to 900 m. **Miscellaneous**: Plant eaten cooked as vegetable, or as a salad. Roots
and stem used in traditional medicine. **Flowering**: September-January.

101. EAST INDIAN GLOBE THISTLE *Sphaeranthus indicus*
Names: Hin, Mar: *Mundi, Gorakh-mundi*. **Plant**: 50 cm. **Flower**: 1-1.5 cm across
Distr.: India, Sri Lanka. **Habit**: An aromatic herb, whose ground hugging branches have
toothed wings. Abundant in open areas in harvested fields and dry pond beds.
Miscellaneous: Small insects pollinate the flowers. Roots, leaves and flowers are used
in traditional medicine. **Flowering**: November-January.

102. MEXICAN SUNFLOWER *Tithonia diversifolia*
Plant: 4 m. **Flower**: 14 cm across. **Distr.**: Native of Mexico, introduced in India. **Habit**:
Now a garden escape, this tall gregarious shrub having alternate divided leaves is seen in
the plains as well as on the hills. At places, it is an aggressive coloniser. Flowers
conspicuous, like sunflowers. **Miscellaneous**: Genus named after Tithonus, a handsome
youth and King of Troy. He was loved by Aurora who turned him into a grasshopper.
Good to grow in herbaceous borders in the background. Cultivated as green-manure
plant. Flower heads used in traditional medicine. **Flowering**: September-May.

97

98

99

100

101

102

103. WESTERN HILL THISTLE *Tricholepis amplexicaulis*
Names: Mar: *Dahan.* **Plant:** 1-2.5 m. **Flower:** 1.5-3 cm across. **Distr.:** Western Ghats (Maharashtra, Karnataka). **Habit:** A stout, erect annual seen mainly on hillsides in small groups. Prefers open hill slopes. Angled, ribbed branches are typical of this group. **Miscellaneous:** Flowers attract bees and butterflies. **Flowering:** November-December.

104. COAT BUTTONS *Tridax procumbens*
Names: Hin: *Akal kohadi,* Tel: *Raavanaasuri di talakai,* Tam: *Vettu-kkaaya-thalai,* Kan: *Gabbu sanna savanthi,* Dharwar: *Tikkikasa.* **Plant:** 30-60 cm. **Flower:** 1-1.5 cm across. **Distr.:** Native of C. America, naturalised in India. **Habit:** A slender perennial hairy herb, abundant along the roads and open areas. This straggling herb is gregarious in habit. **Miscellaneous:** Flowers very attractive to butterflies and bees. **Flowering:** January-December.

105. ASH FLEABANE *Vernonia cinerea*
Names: Sans, Kan: *Sahadevi,* Hin: *Sahadevi, Daudotpala,* Mar: *Sadodi,* Guj: *Sadori,* Beng: *Kukshim,* Tel: *Garitikamma,* Tam: *Mukuttipundu,* Mal: *Puvan kodanthel.* **Plant:** 15-75 cm. **Flower:** 4 mm across. **Distr.:** India, Nepal, Sri Lanka. **Habit:** One of the commonest plants, seen in every possible niche from roof tops by the sea up to the Himalaya (1,800 m). Often variable in appearance. **Miscellaneous:** Genus is named after William Vernon, an English botanist. Leaves, roots and seeds are used in traditional medicine. **Flowering:** January-February.

PLUMBAGINACEAE   Sea-lavender Family

106. WHITE PLUMBAGO *Plumbago zeylanica*
Names: Hin: *Chitrak,* Mar, Kan: *Chitramula,* Guj: *Chitaro,* Beng: *Chitra,* Tel: *Agnimata,* Tam: *Cithiramulam,* Mal: *Thumba,* Ory: *Chitamulo.* **Plant:** 1-1.5 m. **Flower:** 1 cm across. **Distr.:** India, Sri Lanka. **Habit:** A rambling shrub found in scrub jungles, forest edges and fallow land. Tubular calyx is glandular-hairy, and sticky to the touch. **Miscellaneous:** Caterpillars of the Zebra Blue butterfly feed on this plant. Roots and bark used in traditional medicine. **Flowering:** December-April.

PRIMULACEAE   Primrose Family

107. SCARLET PIMPERNEL *Anagallis arvensis*
Names: Hin: *Jonkh-mari,* Guj: *Anagalide,* Punj: *Dhabbar.* **Plant:** 12-50 cm. **Flower:** 8 mm across. **Distr.:** India, Nepal, Sri Lanka. **Habit:** Fairly common herb, but often overlooked, being small. Prefers forested hilly regions. **Miscellaneous:** Used as fish poison and for expelling leeches from the nostrils of cattle. An ingredient in an ecofriendly pesticide. Poisonous to dogs. **Flowering:** August-October.

OLEACEAE   Jasmine Family

108. MALABAR JASMINE *Jasminum malabaricum*
Names: Hin: *Ban mogra,* Mar: *Kundi,* Guj: *Kusuri,* Kan: *Dolle kusdiballi,* Mal: *Tirgal.* **Plant:** 12 m. **Flower:** 3 cm across. **Distr.:** Peninsular India. **Habit:** On lower slopes, it is usually a bushy straggler, but on hills (600-1,000 m) it may be seen on trees as a woody climber. Flowers are fragrant. Genus name derived from Persian *yasmin.* **Miscellaneous:** Seeds are dispersed by birds. Fruit is cooked and eaten locally. **Flowering:** March-May.

103

106

104

107

105

108

109. CHRIST'S THORN *Carissa congesta* (= *C. carandas*)
Names: Sans, Guj: *Kararmarda*, Hin: *Karaunda*, Mar: *Karvanda*, Beng: *Karamcha*,
Tel: *Kalaikkay*, Tam: *Kalakkay*, Kan: *Karekayi*. **Plant**: 5 m. **Flower**: 2 cm across.
Distr.: Warm regions of India (except arid NW), Sri Lanka. **Habit**: Thorny evergreen
shrub with 2-4 cm long thorns, often forked. Flowers have jasmine-like fragrance.
Prefers hill slopes in medium to high rainfall regions. **Miscellaneous**: Seeds are dispersed
by birds. Food plant of Olive Green Hawkmoth (*Nephele didyma*). Fruit edible.
Flowering: March-April.

110. SERPENT ROOT *Rauvolfia serpentina*
Names: Sans, Guj: *Sarpagandha*, Hin: *Chandrabhaga*, Mar: *Harkaya*, Beng: *Chandra*,
Tel: *Paataalagani*, Tam: *Chivan amelpodi*, Kan: *Sarpagandhi*, Mal: *Chuvannavipori*,
Ory: *Patalgarar*. **Plant**: 1.5 m. **Flower**: 5 mm across. **Distr.**: India (absent in arid
regions), Andaman Is., Sri Lanka (now extinct). **Habit**: An erect, evergreen, shade
loving plant, found in moist deciduous and evergreen forests. **Miscellaneous**: Over-
collection has endangered this plant in the wild. The roots are used in medicine.
Flowering: March-May.

111. BLOOD FLOWER *Asclepias curassavica*
Names: Hin: *Kak tund*, Mar: *Kurki*. **Plant**: 1.5 m. **Flower**: 1-1.5 cm across. **Distr.**:
Native of West Indies, naturalised in India, Sri Lanka. **Habit**: A garden escape, which
has successfully naturalised in India. Seen along moist places, more common on hills.
Miscellaneous: Food plant of the Plain Tiger butterfly. Flowers attract butterflies.
Flowering: January-December.

112. GIANT MILKWEED *Calotropis gigantea* Rajasthan '89 onwards – W. Bengal Sikkim '10
Names: Sans: *Arka*, Hin: *Aak, Alarkh*, Mar: *Rui*, Guj: *Akado*, Beng: Akanda, Tel:
Jilledu, Tam: *Arkkam*, Mal: *Erikku*, Sind: *Bijalosha*. **Plant**: 1.5-3 m. **Flower**: 4 cm
across. **Distr.**: India, Sri Lanka. **Habit**: Commonest shrub, particularly in open, arid
regions. Can grow into a small tree. Produces profuse milky latex when bruised.
Miscellaneous: Food plant of the Plain Tiger butterfly and painted grasshopper.
Flowering: January-December.

113. DECCAN CARALLUMA *Caralluma adscendens*
Names: Hin: *Pippa*, Mar: *Makadsing*. **Plant**: 20 cm. **Flower**: 1.5 cm across. **Distr.**:
Peninsular India, Sri Lanka. **Habit**: Gregarious clumps of this succulent, leafless herb
are seen sheltered in thorny bushes. This perennial is locally common in drier regions,
but never abundant. **Miscellaneous**: The putrid smell of the flowers attracts flies for
pollination. Food plant of Plain Tiger butterfly. Plant eaten as a vegetable. **Flowering**:
June-September.

114. FANTASTIC FLYTRAP *Ceropegia fantastica*
Flower: 2.5 cm long. **Distr.**: Western Ghats (Karnataka, Goa). **Habit**: An extremely
rare, endemic twiner of the Western Ghats, prefers open semi-evergreen forests. The
extra long calyx is its unmistakable characteristic. It is one of the most endangered
among Indian plants. **Miscellaneous**: Caterpillars of Striped Tiger and Glassy Tiger
butterflies feed on the plant. **Flowering**: August-September.

115. PEACOCK FLYTRAP *Ceropegia oculata*
Flower: 6.5 cm long. **Distr.**: Western Ghats (Maharashtra). **Habit**: A rare twiner that springs forth every monsoon from the underground tuber, usually sheltered among thorny shrubs on the hill slopes. **Miscellaneous**: Caterpillars of the Striped Tiger and Glassy Tiger butterflies feed on the plant. Extensive collection of tubers for food by tribals, and clearing of forests, has endangered this plant. Tubers are cooked and eaten like potatoes. **Flowering**: July-September.

116. FOREST FLYTRAP *Ceropegia vincaefolia*
Flower: 5-6 cm long. **Distr.**: Western Ghats (Maharashtra). **Habit**: This monsoon twiner grows on the forested slopes, often among thickets of karvi. The tubers of all *Ceropegia* remain dormant till the next rains. Flowers vary in colour and shape. **Miscellaneous**: Caterpillars of the Striped Tiger and Glassy Tiger butterflies feed on the plant. Tribals collect the tubers for food. Generic name is derived from Greek *keros* (wax) and *pege* (a fountain) referring to the waxy flowers. **Flowering**: August-September.

117. WAX-LEAVED CLIMBER *Cryptolepis buchanani*
Names: Mar: *Dudh vel, Karanta*, Tel: *Adavipala tige*. **Flower**: 1.5 cm across. **Distr.**: India, Sri Lanka. **Habit**: A very common perennial twiner with woody base and shining leaves in pairs, opposite to each other. Milky latex oozes if injured. Seen along forest edges. **Miscellaneous**: Caterpillars of the Common Crow butterflies feed on this plant. **Flowering**: March-September.

118. PURPLE TRUMPET *Cryptostegia grandiflora*
Names: Hin: *Vilayti aak*, Mar: *Vilayti vakhandi*, Tam: *Palai*, Mal: *Pala*. **Plant size**: 1.5-3 m. **Flower**: 5 cm across. **Distr.**: Native of Tropical Africa and Madagascar, naturalised in India. **Habit**: A large, evergreen, spreading climbing shrub, seen commonly along the roadsides and fields. This handsome garden escape prefers drier regions of open scrub country. **Miscellaneous**: Tried with partial success as source of rubber. Bark fibre used in making fishing line. **Flowering**: Almost throughout the year.

119. DALZELL'S FREREA *Frerea indica*
Names: Mar: *Makad shingi*. **Plant**: 10-15 cm. **Flower**: 2 cm across. **Distr**: Maharashtra (Deccan hills of Purandar and Junnar, Pune district, Mahabaleshwar, Sajjangad, Satara district, Varandha ghat, Raigad district, Bhandardara, Ahmednagar district). **Habit**: This succulent, perennial herb grows in gravelly soil on exposed bare rocks on hill slopes and cliffs at about 1,000 m. **Miscellaneous**: The plant is often found sheltered within clumps of the thorny *Euphorbia neriifolia*. Star-like flowers attract flies for pollination. Caterpillars of Plain Tiger butterfly feed on the plant. Being extremely rare, and endemic to a very limited area, it has been listed by the IUCN as one of the world's twelve endangered plants. It is listed in Appendix I of the CITES which prohibits its collection and export. **Flowering**: June-October.

120. ROSY MILKWEED TWINER *Oxystelma secamone* (= *O. esculentum*)
Names: Hin, Beng., Ory: *Dudhialata*, Mar: *Dudhani*, Guj: *Jaldudhi*, Tel: *Dudipala*, Tam: *Usippalai*, Kan: *Dugdhike*. **Flower**: 2.5 cm across. **Distr.**: India, Sri Lanka. **Habit**: A branching, slender, perennial twiner. Common on bushes and small trees, especially along canals and water bodies. **Miscellaneous**: Flowers, fruit and leaves eaten during famine. **Flowering**: July-February.

121. PANACEA TWINER *Tylophora indica* (=*T. asthmatica*)
Names: Hin: *Antamul*, Beng: *Antomul*, Mar: *Pitkari*, Guj: *Damni vel*, Tel: *Verripala*, Tam: *Nach-churuppan*, Kan: *Aiitmula*, Mal: *Vallipaala*. **Plant**: 1.5 m. **Flower**: 1 cm across. **Distr.**: India (except arid Northwest), Sri Lanka. **Habit**: Branching twiner, common on thickets. Up to 900 m on hills. **Miscellaneous**: Food plant of Glassy Tiger butterfly. Used in traditional medicine. **Flowering**: February-October.

122. GREEN MILKWEED CLIMBER *Wattakaka volubilis* (=*Dregea volubilis*)
Names: Hin: *Akad bel*, Mar: *Ambri*, Guj: *Malati*, Beng: *Titakunga*, Tel: *Dudhipala*, Tam: *Kodi palai*, Kan: *Dugdhive*, Ory: *Dudghika*. **Plant**: 10 m. **Flower**: 1.5 cm across. **Distr.**: India, Sri Lanka. **Habit**: Perennial, spreading liana seen on rocky slopes, trees and thickets. Commonly found on plains, up to 1,300 m on the hills. **Miscellaneous**: Food plant of Blue Tiger butterfly. **Flowering**: April-May.

MENYANTHACEAE   **Bogbean Family**

123. CRESTED SNOWFLAKES *Nymphoides hydrophylla* (= *N. cristata*)
Names: Hin: *Tagarmul*, Beng: *Panchuli*, Mar: *Khatara, Kolare chikal*, Tel: *Anthara thamara*. **Plant**: 50-85 cm. **Flower**: 2 cm across. **Distr.**: India. **Habit**: A gregarious, aquatic herb with floating leaves, commonly seen in shallow ponds and slow flowing water bodies. Often a dominant feature of a wetland. **Miscellaneous**: Honey bees pollinate the flowers. Leaf stalks and fruits used as vegetable. **Flowering**: January-December.

BORAGINACEAE   **Borage Family**

124. COMMON HILL BORAGE *Adelocaryum coelestium*
Names: Hin: *Sandruk*. **Plant**: 1.5 m. **Flower**: 6 mm across. **Distr.**: Western Ghats (Maharashtra). **Habit**: Erect herb with red branches, grows on forested hills. Common in forest clearings and along paths in hill stations. **Miscellaneous**: Yields an alkaloid essential for the reproduction of milkweed butterflies, which attracts them to bruised plants. **Flowering**: August-November.

125. INDIAN TURNSOLE *Heliotropium indicum*
Names: Hin: *Hatta-juri*, Mar: *Bharundi*, Guj: *Hathi-sundhara*, Beng: *Hati sura*, Tel: *Nagadonti*, Tam: *Nakki-poo*, Kan: *Chalukondee*, Mal: *Thekada*. **Plant**: 75 cm. **Flower**: 3 mm across. **Distr.**: India, Sri Lanka. **Habit**: Aromatic, hairy herb, with a woody base, grows on degraded land. **Miscellaneous**: Yields an essential alkaloid, the plant is attractive to milkweed butterflies, attracts them to dried or bruised plants. Leaves and seeds are used in traditional medicine. Generic name derived from Greek *helios* (the sun) and *trope* (to turn) referring to the old belief that the flowers turned with the sun. **Flowering**: January-December.

126. INDIAN BORAGE *Trichodesma indicum*
Names: Hin, Beng: *Chota kulpha*, Mar: *Chota phulva*, Guj: *Undhaphuli*, Tel: *Guvvagutti*, Tam: *Kazhuthaithumbai*, Kan: *Katte tumbe soppu* Ory: *Hetamundai*. **Plant**: 45 cm. **Flower**: 1 cm across. **Distr.**: India, Nepal, Sri Lanka. **Habit**: Branching, erect annual, common along the roadside around villages and forests. Blue-purple inverted flowers are unmistakable. **Miscellaneous**: Plant attracts milkweed butterflies. **Flowering**: August-October.

CONVOLVULACEAE   Glory Family

127. SILKY ELEPHANT GLORY *Argyreia nervosa*
Names: Sans: *Samudraphalaka*, Hin: *Samundar-ka-pat*, Ghav *bel*, Mar: *Samudra soka*, Beng: *Bichtarak.* **Flower**: 8.5 cm long. **Distr.**: India, absent in drier regions. **Habit**: Large woody, perennial climber with heart shaped leaves having silky, silvery undersides. Large flowers are seen from the latter half of the monsoon. Found in forested regions and on hills up to 900 m. **Miscellaneous**: Tortoiseshell beetles feed on leaves. Carpenter bees pollinate the flowers. Leaves and roots are used in traditional medicine. **Flowering**: July-December and March-April.

128. WHITE GROUND GLORY *Convolvulus microphyllus*
Names: Hin, Raj: *Santari, Santer.* **Plant**: 15-50 cm long. **Flower**: 2 cm across. **Distr.**: Gujarat, Rajasthan. **Habit**: A spreading, slender perennial herb, branching from the base. Seen among grasses and open ground in dry, sandy soil. Flowers are also pinkish-white. **Miscellaneous**: Generic name is derived from Latin *convolva* (to twine around). Plant is used in traditional medicine. **Flowering**: August-December.

129. COMMON DODDER *Cuscuta reflexa*
Names: Hin: *Amar bel*, Mar: *Nirmuli*, Guj: *Akasvel*, Beng: *Haldi-algusilatta*, Tel: *Sitamma pogunalu.* **Flower**: 8 mm long. **Distr.**: India, Sri Lanka, Pakistan. **Habit**: Slender twiner is an aggressive parasite on bushes. Covers in dense interlacing masses. Can be seen in cities as well as forests from coast to hills. **Miscellaneous**: Being a parasite, it is disliked by gardeners as it weakens host plants. Seeds used in traditional medicine. **Flowering**: October-January.

130. LITTLE GLORY *Evolvulus alsinoides*
Names: Sans: *Vishnugandhi*, Hin, Mar, Tel, Kan: *Vishnukranta*, Guj: *Kalisankha vali*, Tam: *Vishnukrandi*, Mal: *Vishnaclandi.* **Plant**: 30 cm. **Flower**: 1 cm across. **Distr.**: India, Sri Lanka. **Habit**: Dainty little spreading, hairy perennial, common on open, grassy slopes. Abundant on hills up to 1,500 m. Flowers are white too. Plant variable in size. **Miscellaneous**: Plant used in traditional medicine. **Flowering**: July-November.

131. MARSH GLORY *Ipomoea aquatica*
Names: Hin, Beng: *Kalmisag*, Mar: *Nadishaka*, Guj: *Nalanibhaji*, Tel: *Tutikura*, Tam: *Ellaikeerai*, Punj: *Ganthian.* **Flower**: 5 cm across. **Distr.**: India, Nepal, Sri Lanka. **Habit**: Common in marshy, water-logged places. A trailing runner with hollow stems that float and root at nodes. **Miscellaneous**: Caterpillars of Death's Head Hawkmoth (*Acherontia lachesis*) and Glory Hawkmoth (*Agrius convolvuli*) feed on the plant. Tender shoots are cooked and eaten. **Flowering**: October-June.

132. RAILWAY GLORY *Ipomoea cairica*
Names: Mar: *Gar vel.* **Flower**: 5 m across. **Distr.**: Introduced in India, Sri Lanka and other tropical countries. **Habit**: Extensively spreading, perennial climbing vine with tuberous roots. Seen on fences and over shrubs and trees, which it covers aggressively. On the hills, it occurs up to 1,000 m. White flowers are also seen. **Miscellaneous**: Commonly planted in gardens and around houses for insulation against summer heat. **Flowering**: January-December.

78

133. **HEDGE GLORY** *Ipomoea carnea*
Names: Hin, Mar: *Besharam*. **Plant**: 2 m. **Flower**: 10 cm across. **Distr.**: Native of S. America, naturalised in India (Himalaya up to 1,400 m), Nepal. **Habit**: An aggressive invader with milky juice and hollow branchlets. Overruns the banks of lakes and ponds, choking out native plants. Gregarious, non-climbing shrub. **Miscellaneous**: Caterpillars of Death's Head Hawkmoth (*Acherontia lachesis*) and two species of Tortoiseshell beetles feed on this plant. Known to absorb heavy metals. Being distasteful to livestock, it is planted as a hedge around houses and farms. **Flowering**: January-December.

134. **RED STAR GLORY** *Ipomoea hederifolia*
Plant size: 8 m. **Flower**: 3-4.5 cm across. **Distr.**: Native of North Mexico and Arizona, naturalised in India. **Habit**: Annual twiner with its bright scarlet flowers that appear as the monsoon trails off. A garden escape spread throughout the tropics. Seen on the hills up to 750 m. **Miscellaneous**: Planted as ornamental in gardens. **Flowering**: October-December.

135. **GREATER GLORY** *Ipomoea mauritiana*
Names: Sans: *Vidari*, Hin: *Bilaikand*, Mar: *Bhuikobola*, Beng: *Bhumikumra*, Tel: *Bhuchakragradda*, Tam: *Palmudangi*, Kan: *Buja-gumbala*, Mal: *Mutalakkanta* **Flower**: 3-6 cm across. **Distr.**: India; absent in drier regions, Sri Lanka. **Habit**: A stout, spreading climber that grows out of large, perennial, tuberous rootstock every monsoon. Seen on shrubs and trees in forested regions. Large showy flowers are conspicuous. **Miscellaneous**: Roots are used in traditional medicine. **Flowering**: July-September.

136. **BLUE DAWN GLORY** *Ipomoea nil*
Names: Sans: *Krishnabija*, Hin, Guj: *Kala dana*, Mar: *Nilpushpi*, Beng: *Nilkalmi*, Tel: *Jirika*, Tam: *Kakkattan*, Kan: *Ganribija*, Punj: *Bildi*. **Flower**: 4-7 cm across. **Distr.**: Throughout India (Himalaya up to 1,800 m). **Habit**: A hairy, annual twiner seen commonly during the latter half of the monsoon along roads and hedges. Flowers open around sunrise and fade before 10 a.m. **Miscellaneous**: Seeds are used in traditional medicine. **Flowering**: August-December.

137. **LESSER GLORY** *Ipomoea obscura*
Names: Hin: *Ker-gawl*, Mar: *Pilibonvari*, Guj: *Gumbadvel*, Tel: *Nallakokkita*, Tam: *Chirudali*, Kan: *Cherutali*. **Flower**: 2.5 cm across. **Distr.**: India, Sri Lanka. **Habit**: Slender, twining annual, common along the roads. On the hills up to 1,400 m. Flowers are pale yellow to yellowish with pink tinge. **Flowering**: August-April.

138. **GOAT'S FOOT GLORY** *Ipomoea pes-caprae*
Names: Hin: *Do-pati lata*, Mar: *Maryadavel*, Guj: *Maryadavela*, Beng: *Chhagalkuri*, Tel: *Bedatige*, Tam: *Adambu*, Kan: *Adum baballi*, Mal: *Atampa*. **Flower**: 6 cm across. **Distr.**: India, other tropical countries. **Habit**: Trailing, sand binding perennial herb with deeply lobed leaves. Though commonly seen along the sandy seashores, it also occurs inland in other sandy habitats. Bilobed leaves give its common and scientific names. **Miscellaneous**: Useful as a sand binder. Leaves are used as vegetable and fodder. **Flowering**: January-December.

139. TIGER'S PAW GLORY *Ipomoea pes-tigridis*
Names: Hin: *Panch patri*, Beng: *Angulilata*, Tel: *Chikunuvvu*, Tam: *Pulichovadi*, Mal: *Pulichevatu*. **Flower:** 4 cm across. **Distr.:** India, Nepal, Sri Lanka. **Habit:** Hairy annual twiner, seen from coast to plains and on the hills up to 900 m. Common and scientific names are given for the shape of the leaves, which varies. Flowers open after 4 p.m. and fade next morning. **Flowering:** September-November.

140. PURPLE HEART GLORY *Ipomoea sepiaria*
Names: Hin, Beng: *Bankalmi*, Mar: *Amtivel*, Guj: *Hanumanvel*, Tel: *Mettatuti*, Tam: *Manjigai*, Mal: *Tirutali*. **Flower:** 3.7 cm across. **Distr.:** India, Sri Lanka. **Habit:** Slender vine seen on bushes. Leaves have purple blotches in the centre. Prefers open areas and exposed slopes on the hills up to 1,200 m. Flowers open after sunrise and close before noon. **Miscellaneous:** Plant is edible and also fed to cattle. Used in traditional medicine. **Flowering:** March-August.

141. EGYPTIAN DAY GLORY *Merremia aegyptia*
Names: Hin: *Ghia bel*, Raj: *Rota bel*. **Flower:** 2.5-3 cm across. **Distr.:** India, drier plains. **Habit:** A silky hairy, slender, annual vine. Commonly seen in the drier regions. Flowers begin to fade by late mid-day. **Flowering:** August-November.

142. WHITE DAY GLORY *Merremia turpethum*
Names: Hin: *Nisoth*, Tel: *Tegada*, Tam: *Sivatai*, Kan: *Sigade*, Mal: *Triputa*. **Flower:** 3 cm across. **Distr.:** India, absent in arid regions. **Habit:** An aggressive invading vine, commonly seen along the road on thickets and trees, covering them extensively. Being a day glory, it blooms long after sunrise (after 9 a.m.) and fades before evening. **Miscellaneous:** The drug Indian Jalap or Turpeth, derived from its roots, is used in traditional medicine. **Flowering:** October-April.

143. GRAPE GLORY *Merremia vitifolia*
Names: Hin: *Nawal ki bel*, Mar: *Navalicha vel*, *Nawal*, Mundari: *Nanrikadsom ba*, Garo: *Dukhumi bidu*. **Flower:** 7 cm across. **Distr.:** India; absent in arid regions; Sri Lanka, Nepal. **Habit:** A perennial, evergreen, hairy vine. Flowers remain open until mid-day. An aggressive vine that takes over large areas, covering other vegetation. Leaves resemble those of grapes, hence both the common and scientific names. **Miscellaneous:** Plant is used in traditional medicine. **Flowering:** October- March.

144. COMMON NIGHT GLORY *Rivea hypocrateriformis*
Names: Hin: *Phang*, Mar: *Kulni*, Beng: *Kalmilata*, Tel: *Boddikura*, Tam: *Budthi kiray*. **Plant:** 8 m. **Flower:** 5 cm across. **Distr.:** India, Nepal. **Habit:** Large, woody vine seen on small bushes and trees. Fragrant, large, showy flowers bloom at twilight and fade around sunrise. Prefers drier scrub forest. Occurs from coast to plains and on the hills up to 900 m. **Miscellaneous:** Young shoots are eaten as vegetable and used in preparing bread. **Flowering:** August-March.

145. THORN APPLE *Datura metel*
Names: Hin, Mar, Guj, Beng: *Dhatura*, Tel, Kan: *Ummattu*, Tam: *Ummattui*. **Plant**: 80 cm. **Flower**: 7 cm across. **Distr.**: India. **Habit**: Common along roadsides. Trumpet-shaped flower has 5 teeth on purple tinged corolla edge. Flowers open by evening and fade before noon. Fruit with fleshy prickles give its common English name. **Miscellaneous**: Highly poisonous. **Flowering**: January-December.

146. ANGEL'S TRUMPET *Datura suaveolens* Sukium 'ro
Names: Hin: *Dhatura*. **Plant**: 4.5 m. **Flower**: 30 cm long. **Distr.**: Native of Tropical America, naturalised in India (Himalayas). **Habit**: Tall shrub with very large, drooping, trumpet shaped white flowers. Common on hills up to 1,700 m. Flowers sweet scented at night. **Miscellaneous**: Flowers attract nocturnal and diurnal insects. Grown as ornamental plant. **Flowering**: March-November.

147. LITTLE GOOSEBERRY *Physalis minima*
Names: Hin: *Pipat*, Mar: *Chirpoti*, Guj: *Parpoti*, Beng: *Ban tapariya*, Tel: *Kupanti*, Tam: *Tholtakkali*, Mal: *Njodinjotta*. **Plant**: 50 cm. **Flower**: 5 mm across. **Distr.**: India, Sri Lanka. **Habit**: Common annual herb along roadsides. On the hills up to 1,000 m. **Miscellaneous**: Fruit edible. Generic name derived from Greek *physa* (a bladder) referring to the bladder-like fruits. **Flowering**: July-February.

148. COMMON INDIAN NIGHTSHADE *Solanum anguivi* (=*S. violaceum, S. indicum*)
Names: Hin: *Barhanta*. **Plant**: 3 m. **Flower**: 2.5 cm across. **Distr.**: India, Nepal, Sri Lanka. **Habit**: Very common, branching, prickly shrub on degraded land. On the hills up to 800 m. Berries orange (0.8 cm across) when ripe. **Miscellaneous**: Food plant of the Death's Head Hawkmoth (*Acherontia lachesis* and *A. styx*). Roots are used in traditional medicine. **Flowering**: August-April.   Nummar 1·13

149. YELLOW-BERRIED NIGHTSHADE *Solanum surattense*
Names: Sans, Beng: *Kantakari*, Hin: *Berkateli*, Mar: *Bhuiringani*, Guj: *Bhoringni*, Tel: *Pinnamulaka*, Tam, Mal: *Kandankathiri*, Punj: *Kandyuli*, Ory: *Bheji-begun*. **Flower**: 2.5 cm across. **Distr.**: India. **Habit**: Perennial shrub, prefers drier open areas. Common along roadsides. Grows close to ground. Branches and leaves armed with sharp prickles. **Miscellaneous**: The root is an important ingredient of the well-known ayurvedic medicine, *Dasamula*. **Flowering**: January-December.

SCROPHULARIACEAE   **Figwort Family**

150. THYME LEAVED GRATIOLA *Bacopa monnieri*
Names: Hin: *Neer Brahmi*, Mar, Tam, Kan, Mal: *Nirbrahmi*, Beng: *Brahmi-sak.* **Flower**: 8 mm across. **Distr.**: India, Nepal, Sri Lanka. **Habit**: Succulent, creeping herb that roots at the nodes, seen in moist places in dense mats. Pale blue or white flowers are short lived. Found on hills up to 1,200 m. **Miscellaneous**: Plant used in traditional medicine. **Flowering**: January-December.

151. COMMON LIMNOPHILA *Limnophila heterophylla*
Plant: 15-20 cm. **Flower**: 0.9 cm long. **Distr.**: India, Sri Lanka. **Habit**: An aquatic herb, often submerged, branching from the base. Lower submerged leaves finely divided, in whorls around the stem, while those above water are long, toothed and arranged opposite on the stem. Grows gregariously in marshes, ditches, ponds and rice-fields. Locally abundant. Generic name derived from Greek *limne* (a marsh) and *philo* (to love). **Flowering**: September-January.

152. SWEET BROOM *Scoparia dulcis*
Names: Santal: *Jastimadhu, Mundari Madhukam, Chinibuta*, Bastar: *Ghoda tulsi, Mithi patti.* **Plant**: 30-90 cm. **Flower**: 5 mm across. **Distr.**: Native of Tropical America, naturalised in India, Nepal, Sri Lanka. **Habit**: An erect annual herb, common in railway yards and roadsides, often gregarious. *Scopa* means broom in Latin, hence the common name. **Flowering**: January-December.

153. COMMON SOPUBIA *Sopubia delphinifolia*
Names: Mar, Guj: *Dudhali*, Santal: *Dak-kadur.* **Plant**: 0.3-1m. **Flower**: 7 mm across. **Distr.**: India, Sri Lanka. **Habit**: Erect annual with grooved stems and finely divided leaves. Common during monsoon on coasts, plains as well as hills up to 1,600 m. Flowers vary from pale pink to pinkish purple. **Miscellaneous**: Root parasite on grasses and other plants. **Flowering**: August-January.

154. PURPLE WITCH *Striga gesnerioides*
Names: Hin: *Missi.* **Plant**: 30 cm. **Flower**: 6 mm across. **Distr.**: Western India (Rajasthan southwards), Sri Lanka, Pakistan. **Habit**: Slender, dark, reddish purple root parasite common under trees and around shrubs. Scale-like reddish purple leaves also serve as floral bracts. More abundant on hills. **Miscellaneous**: Root parasite on the genera *Euphorbia, Lepidagathis.* **Flowering**: August-January.

OROBANCHACEAE   **Broomrape Family**

155. FOREST GHOST FLOWER *Aeginetia indica*
Plant: 25 cm. **Flower**: 2 cm across. **Distr.**: India (Himalaya 600-1,700 m), Nepal, Sri Lanka. **Habit**: Slender, gregarious root parasite seen on shaded forest floor during monsoon. Flowers are red, purple or even white. Typical all-flower leafless plant parasitises the roots of plants. **Miscellaneous**: Plant yields an alcoholic extract, aeginetic acid and a lactone, aeginetolide. **Flowering**: August-October.

156. FOX'S RADISH *Cistanche tubulosa*
Names: Hin, Raj: *Lonki-ka-mula.* **Plant**: 1 m. **Flower**: 5 cm long. **Distr.**: India (Punjab, Rajasthan, Gujarat), Pakistan (Sindh). **Habit**: All-flower bearing, fleshy, root parasite, common in the arid regions and on sandy shores. **Miscellaneous**: Root parasite on *Salvadora persica.* **Flowering**: October-March.

157. DEVIL'S CLAW *Martynia annua*
   Names: Hin: *Hathajori*. **Plant:** 1 m. **Flower:** 2. 5 cm across, 4 cm long. **Distr.:** Native of Mexico, naturalised in India, Sri Lanka, Pakistan. **Habit:** A hairy, branched annual, common on roadsides and rubbish dumps. On the hills up to 900 m. Queer looking clawed fruits give its common English name. **Flowering:** August-October.

158. COMMON PEDALIUM *Pedalium murex*
   Names: Hin: *Bilayati gokhur*, Raj: *Gokhru kanti*. **Plant:** 40-50 cm. **Flower:** 2 cm across, 3 cm long. **Distr.:** Drier regions of India, Sri Lanka, Pakistan. **Habit:** Branching, erect or trailing, annual herb with succulent stem. Prefers drier open regions. A pair of brownish black glands on the stem at each side of the flower. Fruit has four spines. **Flowering:** August-November.

159. ORIENTAL SESAME *Sesamum orientale*
   Names: Sans: *Tila*, Hin, Mar: *Til*, Guj: *Tal*, Tel: *Nuvvulu*, Tam, Kan: *Ellu*, Mal: *Karuthellu*, Punj: *Tili*, Ory: *Khasa*. **Plant:** 75 cm. **Flower:** 1.8 cm across, 2.5 cm long. **Distr.:** India (Himalaya 1,200 m). **Habit:** Bell-like, rose-purple or rarely white flowers adorn this erect hairy annual. Often seen along the roadsides and forest paths, from plains up to 1,200 m on the hills and the coast. **Miscellaneous:** Its cultivated variety is an important oil seed crop. **Flowering:** August-October.

ACANTHACEAE   **Acanthus Family**

160. SEA HOLLY *Acanthus ilicifolius*
   Names: Hin, Beng: *Hargoza*, Mar: *Harkusa*, Tel: *Alasyakampa*, Kan: *Holechudi*, Mal: *Chakkaramulli*. **Plant:** 1.5 m. **Flower:** 3 cm across. **Distr.:** India, Sri Lanka. **Habit:** Gregarious, erect, tall shrub with stiff spiny leaves seen all along the muddy shores of coastal marshes and tidal rivers, associated with mangroves. Plants flower almost gregariously. Locally abundant. **Miscellaneous:** Bees and sunbirds pollinate the flowers. **Flowering:** December-May.

161. COMMON COUGH CURE *Adhatoda zeylanica*
   Names: Sans: *Vasaka*, Hin: *Adusa*, Mar: *Adulsa*, Guj: *Adulso, Ardusi*, Beng: *Bakas*, Tel: *Adasarama*, Tam: *Adadodai*, Kan: *Adusoge*, Mal: *Atalotakam*. **Plant:** 3 m. **Flower:** 1.5 cm across. **Distr.:** India (Himalaya 500-1,600 m) Pakistan, Nepal, Sri Lanka. **Habit:** A tall, aromatic shrub, growing in dense, gregarious clumps. Common hedge plant in villages. **Miscellaneous:** Flowers favoured by bees and sunbirds. Much valued in traditional medicine. Cattle do not eat this plant. **Flowering:** August-November.

162. CREAT *Andrographis paniculata*
   Names: Sans: *Kirata*, Hin: *Kalpanath*, Mar: *Oli-kiryata*, Guj: *Kariyatu*, Beng: *Kalmegh*, Tel: *Nelavemu*, Kan: *Nelaberu*, Mal: *Nelavepu*. **Plant:** 1 m. **Flower:** 5x2.5 mm across. **Distr.:** Peninsular India, Sri Lanka. **Habit:** This slender herb is seen in the plains and coastal regions. Often planted in backyards for medicinal use. **Flowering:** June-September and January-March.

163. VIOLET ASYSTASIA *Asystasia dalzelliana* (= *A. violacea*)
Plant: 1 m. **Flower**: 3 cm across; flower tube: 2.5 cm long. **Distr.**: Western Ghats (Maharashtra, Karnataka). **Habit**: This small shrub is more commonly seen on the hills as undergrowth below trees and large shrubs. It also grows in the moist forests near coastal regions in paartly shaded areas. Often dominant and gregarious. Flowers may be white or blue also. **Flowering**: August-November.

164. CRESTED BARLERIA *Barleria cristata*
Plant: 0.6-2 m. **Flower**: 4 cm across. **Distr.**: India (Himalaya; 600-2,000 m), Pakistan, Nepal, Bhutan. **Habit**: An erect, perennial shrub seen from plains to Himalaya in lightly forested regions. Flowers may occur in purplish pink, violet,or white. Common along forest roads. **Miscellaneous**: Food plant of the Pansy group of butterflies. Often planted in gardens. Leaves and roots used in traditional medicine. **Flowering**: Plains October-March, Himalaya: July-October.

165. FOREST BARLERIA *Barleria prattensis*
Plant: 1 m. **Flower**: 3-5.5 cm across. **Distr.**: Western Ghats. **Habit**: An erect, woody shrub, commonly grows along the forest streams and forest roads. Flowers are seen as the monsoon trails off. **Miscellaneous**: Food plant of Pansy group of butterflies. The genus is named after Jacques Barrelier (1606-1673), French monk and botanist. **Flowering**: August-November.

166. YELLOW HEDGE BARLERIA *Barleria prionitis*
Names: Sans: *Karunta*, Hin: *Jinti , Sahachara*, Mar: *Pivali koranti*, Guj: *Kantashelio*, Beng: *Kanta jati*, Tel: *Mullu goranta*, Tam, Mal: *Shemmuli*, Kan: *Mullu gorante*. **Plant**: 1-1.5 m. **Flower**: 2.5-4 cm across. **Distr.**: India, Sri Lanka. **Habit**: A much branched, prickly shrub throughout warmer parts of India, up to 500 m on the hills. Seen commonly along roadsides. **Miscellaneous**: Caterpillars of the Pansy group of butterflies feed on this plant. Thorny, often planted as hedge. Roots used in traditional medicine. **Flowering**: November-May.

167. HILL BLEPHARIS *Blepharis asperrima*
Flower: 2.5 cm long. **Distr.**: Western Ghats and hills in Deccan. **Habit**: A low growing, creeping or semi-erect, much branched, hairy shrub which roots at the nodes. At places flowers may be white. Seen commonly on hill slopes and in forests. Endemic to the region. **Miscellaneous**: Bees pollinate the flowers. **Flowering**: November-April.

168. CREEPING BLEPHARIS *Blepharis maderaspatensis*
Flower: 1.2x0.3 cm across. **Distr.**: India, Sri Lanka. **Habit**: Prostrate, creeping, wiry plant, rooting at the nodes. Seen commonly on slopes among rocks, poor gravelly soil on the hills up to 1,400 m. **Miscellaneous**: Bees pollinate the flowers. **Flowering**: November-February.

169. COMMON CONEHEAD / KARVI *Carvia callosa*

Names: Hin: *Maraudana*, Mar: *Karvi*, Guj: *Pandadil*. **Plant**: 3 m. **Flower**: 4 cm across. **Distr.**: Western Ghats (Gujarat, Maharashtra, north Karnataka), central India (Pachmarhi) and Mount Abu. **Habit**: Gregarious, erect shrub, grows along the hilly slopes. Mass flowers every eight years, though some patches flower intermittently every year. **Miscellaneous**: Food plant of Blue Oakleaf, Chocolate Pansy and Malabar Spotted Flat butterflies. Mass flowering brings exceptionally high yield of dark amber honey. **Flowering**: August-November every eight years.

170. BLUE ERANTHEMUM *Eranthemum roseum*

Names: Hin: *Nil vasak*, Mar: *Dasamuli*, Tam: *Nilamulli*. **Plant**: 30-200 cm. **Flower**: 2.5 cm across; flower spike: 15 cm long. **Distr.**: Western Ghats (Maharashtra, Karnataka). **Habit**: Common on the forest floor under large shrubs and trees. Tall flower spike is unmistakable, showing white bracts with raised green nerves. Flowers which fade to red, emit unpleasant odour. **Flowering**: November-April.

171. SPINY BOTTLE-BRUSH *Haplanthodes verticillatus*

Names: Hin: *Kastula*, Mar: *Jhankara*, West Indian: *Kalakiryat*. **Plant**: 45-75 cm. **Flower**: Corolla 1.5 cm long. **Distr.**: Western Ghats (Maharashtra, Karnataka). **Habit**: An erect herb of forested hills, seen sheltered under large shrubs and trees. Flowers present among hairy spines, each of which has two sharp points. **Miscellaneous**: Source of nectar for insects. Plant used in traditional medicine. **Flowering**: December-January.

172. MARSH BARBEL *Hygrophila schulli* (= *H. auriculata*)

Names: Hin: *Talimkhana*. **Plant**: 80 cm. **Flower**: 1.5x0.5 cm across. **Distr.**: India, Nepal, Sri Lanka. **Habit**: Gregarious stands of this thorny, robust shrub are common in and around freshwater marshes and ponds. Though uncommon on hills, it is seen up to 1,400 m in the Himalayan foothills. **Miscellaneous**: Caterpillars of the Pansy group of butterflies feed on this plant. Generic name derived from Greek *hygros* (moist) and *philos* (loving). Seeds are used in traditional medicine. **Flowering**: October-April.

173. MARSH CARPET *Hygrophila serpyllum*

Names: Hin, Guj: *Sarpat*, Mar: *Ran-tewan*. **Plant**: 10-35 cm. **Flower**: 6 mm across. **Distr.**: Bihar, central India and Western Ghats. **Habit**: A creeping herb, often seen in dense clusters carpeting the moist open ground; at times in sheltered situations it is low growing but erect. Usually overlooked, but it is conspicuous when mass flowering. **Miscellaneous**: Source of nectar for rock bees, small butterflies. Leaves eaten during scarcity of food. **Flowering**: September-January.

174. HAIRY CREAT *Indoneesiella echioides*

Plant: 20-50 cm. **Flower**: 8 x 3 mm across. **Distr.**: India, Sri Lanka, Pakistan. **Habit**: This herb, with symmetrical branches on either side, is unmistakable. Often gregarious, it is common in drier regions in open scrub, among sheltered rock piles or on field borders. **Miscellaneous**: Bees pollinate the flowers. **Flowering**: August-March.

175. HILL JUSTICIA *Justicia betonica*
Names: Hin: *Hadpat, Mokander,* Tel: *Tellarantu,* Tam: *Velimungil,* Mal: *Vellakurunji,*
Plant: 2 m. **Flower**: 8x5 mm across. **Distr.**: India, Sri Lanka. **Habit**: This gregarious
annual herb, bearing leafy white bracts with green nerves, is common on open hill
slopes and rocky ravines up to 900 m. The genus is named after James Justice, 18th
century Scottish botanist. **Miscellaneous**: Bees prefer these flowers. The plant is
used in traditional medicine. **Flowering**: July-September.

176. COMMON SMALL JUSTICIA *Justicia procumbens*
Names: Sans: *Kalmashi,* Hin: *Makhania ghas,* Raj: *Kagner,* Mar: *Karambal,* Tel:
Pitpapada, Tam: *Ottu pillu,* Kan: *Palkodi,* Mal: *Nerel-poottie.* **Plant**: 10-15 cm. **Flower**:
6 mm across. **Distr.**: Throughout India, Nepal, Sri Lanka, Pakistan. **Habit**: A small,
gregarious, annual herb, common on plains, hill slopes and clearings in forests. Grows
on the hills up to 1,200 m. **Miscellaneous**: Flowers attract bees and butterflies.
Flowering: July-September.

177. PIN-CUSHION *Neuracanthus sphaerostachys*
Plant size: 45-60 cm. **Flower**: 1.5 cm across. **Distr.**: Western Ghats up to Gujarat,
Deccan. **Habit**: Stands of this erect, gregarious shrub are common in the deciduous and
mixed forests. Depending on the moisture available, this shrub is annual or perennial.
Miscellaneous: Butterflies and moths favour these flowers. Roots of the plant are
used in traditional medicine. **Flowering**: August- October.

178. HILL CONEHEAD *Nilgirianthus reticulata*
Names: Hin: *Birkubat,* Mar: *Banchimani,* Mal, Tam: *Gopuram thangi,* Guj: *Kalu
kariyatu.* **Plant**: 2 m. **Flower**: 2.5 cm across. **Distr.**: Peninsular India. **Habit**: This erect
shrub occurs exclusively on hills above 1,200 m. Though not as gregarious as karvi, it
flowers after certain periodic gaps. **Miscellaneous**: Bees pollinate the flowers.
Flowering: August-October (not every year).

179. FALSE BARLERIA *Petalidium barlerioides*
Plant: 0.6-1.2 m. **Flower**: 3.8 cm long. **Distr.**: India (absent in arid regions). **Habit**:
Gregarious stands of this shrubby perennial are often seen along forest roads. Large
white flowers have leaf-like, prominently veined secondary bracts with serrated margin,
which are green initially, later turning pale brown. **Flowering**: March-May.

180. WAYSIDE TUBEROSE *Ruellia tuberosa*
Names: Hin: *Chatakni phali,* Tel: *Chetapatakaayala mokka,* Tam: *Tapas kaaya,*
Mundari: *Ote sirka ba.* **Plant**: 45-75 cm. **Flower**: 4 cm across. **Distr.**: Native of
Tropical America, naturalised in India, Sri Lanka. **Habit**: This erect, annual herb is
common in untended gardens and along roads. Seen on the hills up to 1,000 m. **Flowering**:
Throughout the year.

181. **BLACK-EYED SUSAN VINE** *Thunbergia alata*
Flower: 1.8 cm across. **Distr.**: Native of Tropical Africa, naturalised in India. **Habit**: A twiner, often seen among hedges as well as on the ground. Commonly grown in gardens and has naturalised as an escape. More common around the Deccan plains and hills. **Miscellaneous**: Leaf paste is used in traditional medicine. **Flowering**: October-November.

182. **COMMON FOREST THUNBERGIA** *Thunbergia fragrans*
Names: Kan: *Indrapushpaballi*, Mal: *Noorvan valli*, Mar: *Chimine*. **Flower**: 2.5x1.8 cm. **Distr.**: India (including Himalaya, Sikkim 500-2,600 m), central Nepal, Sri Lanka. **Habit**: Slender, perennial twiner, common along hill roads in moist, wooded areas. Flowers are not fragrant. **Miscellaneous**: The genus is named after Carl Peter Thunberg (1743-1828), Dutch physician and botanist. Cultivated as a garden plant. **Flowering**: August-November.

183. **MYSORE THUNBERGIA** *Thunbergia mysorensis*
Flower: 8 cm long. **Distr.**: Western Ghats (N. Karnataka and Nilgiris, Tamil Nadu). **Habit**: A rare climber found in the evergreen, forested foothills of the Western Ghats. It is more often seen in gardens, especially in the hills and the Deccan plains grown for its unusually showy flowers. **Flowering**: October-January.

VERBENACEAE   **Vervain Family**

184. **TURK'S TURBAN** *Clerodendrum indicum* (= *C. siphonanthus*)
Names: Sans: *Bhargi*, Hin: *Bharang*, Tel: *Bharangi*, Tam: *Kavalai*. **Plant**: 1.2-2.5 m. **Flower**: 10 cm long. **Distr.**: India (Himalaya up to 1,400 m; Sikkim eastwards), Western Nepal, Bhutan. **Habit**: A tall shrub with long, narrow, pointed, oleander-like leaves. Seen throughout India as a garden escape along roadsides, cultivated areas and near human dwellings. **Miscellaneous**: Caterpillars of the Common Silverline (*Spindasis vulcanus*) butterfly and Death's Head Hawkmoth (*Acherontia styx*) feed on the plant. Root and leaves are used in traditional medicine. **Flowering**: January-December.

185. **COMMON HEDGE BOWER** *Clerodendrum inerme*
Names: Sans: *Kundali*, Hin: *Sankupi*, Mar: *Vanajai*, Beng: *Banjai*, Tel: *Takkolakamu*, Tam: *Anjali*, Kan: *Kundali*, Mal: *Nirnochi*. **Plant**: 0.9-3 m. **Flower**: 2 cm across. **Distr.**: India (coastal), Sri Lanka. **Habit**: Spreading shrub, mainly along the coast above high-tide mark, as a mangrove associate, and spread inland as it makes an excellent hedge. **Miscellaneous**: Caterpillars of Death's Head Hawkmoth (*Acherontia styx*) and Common Silverline butterfly (*Spindasis vulcanus*) feed on the plant. Flowers are a good source of nectar for butterflies, hawkmoths and sunbirds. Provides secure shelter for nesting birds; little bees (*Apis florea*) prefer to set up their colony in this shrub. Bitter root and stem are used in traditional medicine. **Flowering**: November-January.

186. **BLUE FOUNTAIN BUSH** *Clerodendrum serratum*
Name: Hin, Sans: *Bharangi*, Mar, Guj: *Bharungi*, Tel, Kan: *Gantubarangi*, Tam: *Angarvalli*, Mal: *Cherutekku*. **Plant**: 0.9-4 m. **Flower**: 2 cm across. **Distr.**: India, Sri Lanka. **Habit**: Prefers hilly forested terrain up to 1,600 m, where it occurs as small or large shrubs. **Miscellaneous**: Flowers attract butterflies. Roots and leaves are used in traditional medicine. Flowers eaten locally as vegetable. The generic name is derived from Greek *kleros* (chance) and *dendron* (a tree) referring to its variable medicinal properties. **Flowering**: August-September.

187. HILL CLERODENDRUM *Clerodendrum viscosum* (= *C. infortunatum*)
Names: Hin, Beng: *Bhant*, Mar: *Bhandira*, Tel: *Gurrapukattiyaku*, Tam: *Karukanni*, Kan: *Basavanapada*, Mal: *Peruku*. **Plant:** 1-2 m. **Flower:** 1.6 cm long. **Distr.:** India (Himalaya up to 1,500 m., Uttar Pradesh to Sikkim), Nepal, Bhutan. **Habit:** An aggressive coloniser, more common on hills and in forest clearings. Pink-centred white flowers are sweet scented in the evening, but odourless during the day. **Miscellaneous:** Tubular flowers pollinated by long-tongued hawkmoths. **Flowering:** September-April.

188. COMMON LANTANA *Lantana camara* Eveywhere, always!
Names: Hin: *Raimuniya*, Mar: *Ghaneri*, Tel: *Pulikampa*, Tam: *Unnichedi*, Kan: *Natahu-gida*, Mal: *Amipu*. **Plant:** 1.0-2.25 m. **Flower:** 4 mm across; flower head: 4 cm across. **Distr.:** Native of tropical America, naturalised throughout India (Himalaya up to 1,500 m). **Habit:** Widespread invader, this plant has taken over large tracts of land. A scrambling, evergreen, strong smelling shrub with stout recurved prickles. **Miscellaneous:** Birds disperse the seeds. Flowers attract butterflies and moths. Food plant of Death's Head Hawkmoth (*Acherontia styx*). Used in traditional medicine. **Flowering:** January-December.

189. BANK MAT *Phyla nodiflora*
Names: Hin, Beng: *Bhui okra, Jalpapli*, Mar: *Ratoliya*, Guj: *Ratveliyo*, Tel: *Bokenaku*, Tam: *Poduthalai*, Kan: *Nela-hippali*, Mal: *Kattu-thippali*, Ory: *Bukkan*. **Plant:** 30-90 cm. **Flower:** 2.5 mm across; flower spike: 1-2 cm long. **Distr.:** India, Nepal, Sri Lanka. **Habit:** A perennial, creeping herb with very small, pinkish white flowers on stalked spikes. Compact mats of this plant can be seen along moist banks of ponds, ditches and streams. Occurs mainly from coasts up to 500 m in the hills. **Miscellaneous:** Flowers attract small butterflies. **Flowering:** January-December.

190. JAMAICAN BLUESPIKE *Stachytarpheta jamaicensis*
Names: Hin: *Kariyartharani*. **Plant:** 50-75 cm. **Flower:** 0.8-1 cm across; flower spike: 10-30 cm long. **Distr.:** Native of Tropical America, naturalised in India, Sri Lanka. **Habit:** Gregarious and common along stream banks, roadsides and degraded habitats. Prefers forests on the plains and on hills up to 1,400 m. **Miscellaneous:** Flowers are very attractive to butterflies. Food plant of Death's Head Hawkmoth. **Flowering:** October-April.

191. COMMON CHASTE TREE *Vitex negundo*
Names: Hin: *Sambhalu*, Mar: *Nirgudi*, Guj: *Nagod*, Beng: *Nisinda*, Tel: *Vaavli*, Tam: *Vellai-nocohi*, Kan: *Lakhigida*, Mal: *Vellanocchi*, Ory: *Beyguna*. **Plant:** 4-6 m. **Flower:** 7 mm across. **Distr.:** India (including Himalaya up to 2,000 m), Nepal, Bhutan, Sri Lanka, Pakistan. **Habit:** Common shrub along roadside as hedges and on river banks. Occurs from coast to plains and up to 1,000 m in the hills. **Miscellaneous:** Flowers attract butterflies and moths. Food plant of the Death's Head Hawkmoth. Leaves repel stored grain pests. Used in traditional medicine. **Flowering:** January-May, July-October.

LABIATAE (LAMINACEAE)   Mint Family

192. WESTERN HILL CATMINT *Anisomeles heyneana*
Names: Hin, Mar: *Chandhara*. **Plant:** 0.6-1.5 m. **Flower:** 1.8 cm across. **Distr.:** Western Ghats. **Habit:** A hairy annual herb, common at forest edge and roadside, but never abundant. Seen from coast to hills. **Miscellaneous:** Flowers are pollinated by bees, wasps and ants. **Flowering:** October-November.

193. INDIAN CATMINT *Anisomeles indica* (= *A. ovata*)
Names: Mar: *Chodhara*, Tel: *Moga-biraku*, Tam: *Peyamerathi*, Kan: *Karitumbe*, Mal: *Karithumba*. **Plant**: 1.5-2 m. **Flower**: 1.5 x 0.4 cm across; flower spike: 7-16 cm. **Distr.**: India (including Himalaya from 200-2,400 m; Uttar Pradesh-Sikkim), Nepal, Sri Lanka. **Habit**: A perennial, bushy shrub that grows in clumps from rootstock or seeds. Seen in disturbed and undisturbed areas, from coast to hills. The plant is scarcely aromatic. **Miscellaneous**: Flowers are pollinated by nectar gathering bees, wasps, ants and sunbirds. Caterpillars of Death's Head Hawkmoth feed on the plant. It is used in traditional medicine. **Flowering**: September-December.

194. INDIAN SQUIRREL TAIL *Colebrookea oppositifolia*
Names: Hin: *Binda, Pansra*, Mar: *Bhaman*. **Plant**: 4-6 m. **Flower**: 0.3x0.2 mm across; flower spike: 5-10 cm long. **Distr.**: India (including Himalaya up to 1,700 m; Kashmir to Arunachal Pradesh), Pakistan, Nepal, Bhutan. **Habit**: A much-branched shrub with woolly young shoots. Gregarious clumps are very common on the hills. **Miscellaneous**: Caterpillars of Death's Head Hawkmoth feed on these plants. Leaves are used in traditional medicine. **Flowering**: December-March.

195. BEARDED MARSH-STAR *Dysophylla stellata*
Plant: 7-20 cm. **Flower**: 2 mm across. **Distr.**: Hills of the Western Ghats. **Habit**: A very pretty, slender herb, which grows gregariously in the shallow pools of stagnant waters, mainly in the hills. Usually seen around receding lake shores, drying pond beds and rice fields. **Miscellaneous**: Source of nectar for butterflies, bees and other insects. **Flowering**: September-December.

196. AMERICAN MINT *Hyptis suaveolens*
Names: Hin: *Vilayti tulsi*, Mar: *Jungli tulas*, Beng: *Bilaiti tulsi*, Bihar: *Bhunsuri*, Ory: *Ganga tulsi*. **Plant**: 0.6-2 m. **Flower**: 5x2 mm across. **Distr.**: Native of Tropical America and West Indies, naturalised in India. **Habit**: Widespread invader, strongly aromatic, seen along country roads, degraded land and forest paths. **Miscellaneous**: Flowers are a favourite with butterflies as a source of nectar. Predators like crab spiders, lynx spiders and mantids lie in wait on flowers for insects. **Flowering**: October-February.

Munnar 1·13

197. LION'S EAR *Leonotis nepetiifolia*
Names: Hin, Beng: *Hejurchei*, Mar: *Dipmal*, Guj: *Matijer, matisul*, Tel: *Ranabheri*. **Plant**: 1.5-3 m. **Flower**: 1.5 x 0.4 cm across, tube 1cm long. **Distr.**: Native of tropical Africa, naturalised in India. **Habit**: An erect, tall, stout, annual herb. Common in degraded land and on roadsides. Grows gregariously, often dominant. **Miscellaneous**: Bees and sunbirds pollinate the flowers. Generic name *Leonotis* is derived from *leon* (lion) and *otos* (ear), because of shape of the flower. **Flowering**: December-March.

198. COMMON LEUCAS *Leucas aspera*
Names: Hin: *Gopha, Chota halkkusa*, Mar: *Tamba*, Tel: *Tummachettu*, Tam: *Thumbai*, Kan: *Thumbe gida*, Mal: *Thumba*, Ory: *Bhutamari*. **Plant**: 15-50 cm. **Flower**: 1.1 cm long. **Distr.**: India. **Habit**: An erect, slender, fragrant annual, which grows gregariously, and is seen commonly in harvested fields and on roadsides. **Miscellaneous**: Bees pollinate the flowers. Leaves are used in traditional medicine. Used as pot herb. **Flowering**: July-November.

193

196

194

197

195

198

NYCTAGINACEAE — Four o'Clock Family

199. COMMON HOGWEED *Boerhavia diffusa*
Names: Hin: *Gadahpurna*, Beng, Tel: *Punarnava*, Mar: *Rakta punarnava*, Guj: *Satodi*, *Vakha khaparo*, Tam: *Mukaratte-kirei*. **Plant**: 1 m. **Flower**: 4 mm across. **Distr.**: India, Sri Lanka. **Habit**: Perennial, spreading herb, abundant in degraded land. Common from coast to hills, up to 1,000 m. **Miscellaneous**: Food plant of the Hogweed Hawkmoth (*Hippotion boerhaviae*). Used in traditional medicine and as a pot herb. **Flowering**: July-September.

AMARANTHACEAE — Cockscomb Family

200. TANGLE MAT *Alternanthera sessilis*
Names: Ben. *Jaljambo*, Hin: *Gudrisag*, Mar: *Kachri*, Tel: *Ponnagunta kura*, Kan: *Honagone soppu*. **Plant**: 50 cm. **Flower**: 1.5 mm across. **Distr.**: India, Nepal, Sri Lanka. **Habit**: Common, low growing herb roots at the nodes. Leaf size and shape is variable. Occurs in dense mats. Seen from coast to plains and on the hills up to 1,500 m. **Miscellaneous**: Flowers attract insects. Used in traditional medicine. **Flowering**: January-December.

201. PRICKLY AMARANTH *Amaranthus spinosus*
Names: Sans: *Tanduliya*, Hin: *Kanta chaulai*, Mar: *Kante math*, Guj: *Kantanu-dant*, Beng: *Kantanotya*, Tel: *Mullutota-kura*, Tam: *Mulluk-kirai*, Kan: *Mullu harive soppu*, Mal: *Kattu-mullen-keera*. **Plant**: 0.3-1 m. **Flower**: 1.5 mm across. **Distr.**: India, Nepal, Sri Lanka (presumably a native of America). **Habit**: Usually annual, often gregarious, this herb is common near rubbish dumps. Small flowers are clustered in dense pendant spikes. It is the only prickly species among the Amaranths. Seen on the hills up to 1,500 m. **Miscellaneous**: Tender leaves and shoots are cooked and eaten. **Flowering**: January-December.

202. SILVER SPIKED COCKSCOMB *Celosia argentea*
Names: Hin: *Safed murga*, Mar: *Kurdu*, Guj: *Lapadi*, Tel: *Panchechettu*. **Plant**: 1 m. **Flower**: 4 mm across; flower spike: 8-14 cm. **Distr.**: India, Nepal, Sri Lanka. **Habit**: Erect, gregarious annual, common on hilly terrain. Often dominant. **Miscellaneous**: Spikes attract butterflies. Tender shoots are cooked and eaten. **Flowering**: August-December.

POLYGONACEAE — Buckwheat Family

203. COMMON MARSH BUCKWHEAT *Polygonum glabrum (=Persicaria glabra)*
Names: Mar: *Sheral*, Beng: *Bihagni*, Tam: *Actalaree*, Kan: *Niru kanigalu*, Mal: *Chavanna mudela mukkum*. **Plant**: 1.5 m. **Distr.**: India, Nepal, Sri Lanka. **Habit**: Annual herb seen growing gregariously in marshy areas, ascending to 1,900 m on the hills. **Miscellaneous**: Flowers attract butterflies. Pungent young shoots are cooked with other vegetables. **Flowering**: September-May.

ARISTOLOCHIACEAE — Birthwort Family

204. GROUND BIRTHWORT *Aristolochia bracteolata (=A. bracteata)*
Names: Hin: *Kiramar*, Raj: *Hukka-bel*, Mar: *Kidemar*, Guj: *Kidamari*, Tel: *Gadidha-gadappa*, Tam, Mal: *Aduthinapalai*, Kan: *Adumuttada-gida*. **Plant**: 50 cm. **Flower**: 5 cm long. **Distr.**: India, Sri Lanka, Pakistan. **Habit**: Low growing herb, more common in open, drier regions. Seen around cultivated fields. **Miscellaneous**: Food plant of the Common Rose (*Pachliopta aristolochiae*). This bitter plant is used in traditional medicine. **Flowering**: August-February.

199

202

200

203

201

204

THYMELEACEAE   **Daphne Family**

205. FISH-POISON BUSH *Gnidia eriocephalus*
Names: Mar: *Ramita, Rametta, Rami*, Tam: *Nachinaar*, Kan: *Enujarige*, Mal: *Nangu*.
Plant: 3.0 m. **Flower**: 3.8 cm across. **Distr.**: Hills of Deccan, Western Ghats (Maharashtra
to Kerala, Tamil Nadu), Sri Lanka. **Habit**: Hardy shrub, common in open forests on the
hills up to 2,100 m. **Miscellaneous**: Plant can cause dermatitis. Stem and leaves used
to stun fish. Leaves are used in traditional medicine. **Flowering**: November-January.

LORANTHACEAE   **Mistletoe Family**

206. LONG-LEAVED MISTLETOE *Dendropthoe falcata*
Names: Hin: *Banda*, Mar: *Vanda*, Guj: *Vando*, Beng: *Baramanda*, Tel, Kan: *Badanika*,
Tam: *Plavithil*, Mal: *Ithil*, Punj: *Amut*. **Flower**: 2.5-5 cm long. **Distr.**: India (absent in
drier regions) (Himalaya: Uttar Pradesh to Sikkim), Bhutan, Sri Lanka. **Habit**: Long,
leathery leaves with reddish midrib, that do not resemble those of the host tree, make it
easier to identify this partial parasite on a variety of host trees. Seen on mangroves on
the coast, and up to 1,500 m on the hills. **Miscellaneous**: Food plant of Jezebel,
Gaudy Baron, White Royal, Peacock Royal, Straight Line Royal. Pollinated by sunbirds.
The sticky seeds are dispersed by birds. **Flowering**: October-February.

EUPHORBIACEAE   **Spurge Family**

207. ANNUAL POINSETTIA *Euphorbia cyathophora* (= *E. heterophylla*)
Names: Hin: *Titli phool*. **Plant**: 2 m. **Flower**: 4x3 mm across. **Distr.**: Native to eastern
United States and Mexico, naturalised in India. **Habit**: Annual shrub, probably a garden
escape, it is now widespread and seen from plains to hills up to 500 m. Upper floral
leaves are green towards the apex, crimson or rose-coloured at the base. **Flowering**:
August-March.

208. LESSER GREEN POINSETTIA *Euphorbia heterophylla* (= *E. geniculata*)
Plant: 10-75 cm. **Flower**: 3x3 mm across. **Distr.**: Native of Tropical America, naturalised
in India. **Habit**: An erect annual herb, often seen in sheltered, untended gardens and
fallow fields growing gregariously. Upper floral leaves form a green rosette. Seen on the
hills up to 800 m. Prefers drier rain shadow regions. This alien invader often forms a
dominant undergrowth, discouraging other plants. **Flowering**: January-December.

209. COMMON SPURGE *Euphorbia hirta*
Names: Hin: *Dudhi*, Raj: *Dhedi-dudheli*. **Plant**: 15-50 cm. **Flower**: 7x6 mm across.
Distr.: Warmer regions of India, Nepal, Sri Lanka. **Habit**: Common annual herb seen
along roadsides. Leaves vary from green to coppery-red, depending on its habitat. Seen
from coast to plains, up to 1,400 m on the hills. **Miscellaneous**: Used in traditional
medicine and eaten as vegetable. **Flowering**: January-December.

210. COMMON HILL SPURGE *Euphorbia rothiana*
Names: Mar: *Dudhi*. **Plant**: 1 m. **Flower**: 2.5x 3 mm across. **Distr.**: Peninsular and
Central India, Sri Lanka. **Habit**: An annual and at times perennial, this erect, branching
shrub is common on the hills up to 1,200 m. **Flowering**: August-November.

205

206

207

208

209

210

211. BANDED BUTTON ORCHID *Acampe praemorsa*
Names: Mar: *Kanbher*, Kan: *Mazrabale*, Mal: *Taliyamaravazha*. **Plant**: 30 cm. **Flower**: 8 mm across. **Distr.**: Peninsular India up to West Bengal, Sri Lanka. **Habit**: This epiphytic orchid is seen in clumps on tree branches and sometimes among crevices of rocks. Common in forested regions from coast to hills up to 1,000 m. **Miscellaneous**: Used in traditional medicine. **Flowering**: April-August.

212. SMALL FLOWERED VANDA *Vanda testacea*
Names: Tel. Vajnika. **Plant**: 10-30 cm. **Flower**: 1.5cm (flower spike 17 cm). **Distr.**: India (Central and Eastern Himalaya to southwards up to Kerala.), Nepal, Bhutan, Sri Lanka, Myanmar. **Habit**: A small to medium sized epiphyte. Erect flower spike has 5 to 20 inconspicuous, fragrant flowers. Seen in broad-leafed, moist and dry deciduous forests in the plains up to 1200 m in warm to hot regions of the subcontinent. **Flowering**: April – July.

213. COMMON SPURRED DENDROBIUM *Dendrobium barbatulum*
Names: Mar: *Jadhia-lasan*. **Plant**: 15-40 cm. **Flower**: 1.5-2 cm across. **Distr.**: Western Ghats (Gujarat, Maharashtra up to Kerala), Deccan hills. **Habit**: This epiphyte, with long tapering stems, is leafless when flowering. It grows abundantly in deciduous forested tracts of the Western Ghats. **Miscellaneous**: Rapidly losing its habitat with forest degradation and collection for ornamental purposes. *Dendrobium* is derived from Greek *dendron* (a tree) and *bios* (life) referring to its epiphytic habit. **Flowering**: April-May.

214. LONG-TAILED HABENARIA *Habenaria commelinifolia*
Plant: 60-90 cm. **Flower**: 2.5-5.6 cm long. **Distr.**: Western Ghats, Central India, W. Bengal, Orissa, Bihar, (Himalaya; Punjab to Kumaon, Arunachal Pradesh), Nepal. **Habit**: Tall flowering stalk of this ground orchid stands out conspicuously among the green monsoon vegetation. Flowers are not fragrant. Seen on the hills up to 2,000 m. **Miscellaneous**: Tuber collection by locals is affecting plant population. **Flowering**: August-September.

215. SINGLE-LEAVED HABENARIA *Habenaria grandifloriformis*
Plant: 7.5-20 cm. **Flower**: Flower spike: 7.6-20 cm tall. **Distr.**: Western Ghats, Deccan Hills. **Habit**: This is among the first flowering herbs to appear with the onset of monsoon. Often seen growing gregariously, carpeting the grassy slopes up to 1,600 m on the hills. As the name suggests, the single, waxy, broad, heart-shaped leaf, flat near the base, is unmistakable. Locally abundant. **Miscellaneous**: Destructive tuber collection is harming the species. **Flowering**: June-July.

216. FOX BRUSH ORCHID *Aerides maculosum*
Names: *Thipke irid amri* [Marathi], *Drupadi Pushpa* [Kannada]. **Plant**: 2.5 -8 cm [stem]. **Flower**: 2 cm (flower spike 35cm long). **Distr.**: Western Ghats,and Eastern Ghats [AP & Orissa]. **Habit**: Grows in moist deciduous as well as semi-evergreen forests up to 1200 m. Fragrant flowers lasts for more than a month on arching, pendulous, 35cm long spike. Flowers are purple, pink or white. Listed as Threatened Species.
Flowering: May-June

211

214

212

215

213

216

217. CHECKERED VANDA *Vanda tessellata*
Names: Sans: *Atirasa*, Hin: *Banda,Vanda*, Mar, Beng: *Rasna*, Guj: *Rasno, Tel*: *Chuttiveduri*, Kan: *Vandakigidda*. **Plant**: 60 cm. **Flower**: 3.5-5 cm across. **Distr.**: India (Gujarat, South India, Uttar Pradesh, Madhya Pradesh, Orissa, W. Bengal, Assam), Sri Lanka. **Habit**: An epiphyte with very showy flowers that are pale violet on opening, turning brown before fading. Locally abundant, prefers deciduous forests. Seen in the forested regions from plains up to 750 m on the hills. This epiphyte prefers mango and *Terminalia* trees. **Miscellaneous**: Generic name derived from Hindi. Used in traditional medicine, being destroyed by root collection. Flowering: May-July.

ZINGIBERACEAE   **Ginger Family**
218. SPIRAL GINGER *Costus speciosus*
Names: Hin: *Kebu*, Mar: *Kosht, Penva*, Beng: *Keu*, Tel, Kan: *Chengalvakoshtu*, Tam: *Kuiravam*. **Plant**: 1.5 m. **Flower**: 4-5 cm across. **Distr.**: India (absent in drier regions), Sri Lanka. **Habit**: Gregarious herb seen along sheltered gravelly slopes in the warm, moist, forested regions. This succulent herb has leaves growing spirally around the stem. **Miscellaneous**: Bitter rhizome used in traditional medicine. **Flowering**: August-October.

219. HILL TURMERIC *Cucurma pseudomontana*
Names: Hin: *Kachura*, Mar: *Sindarbar, Shindalvan*, Guj: *Kachuri*. **Plant**: 65 cm. **Flower**: Flower spike: 5-12.7 cm long. **Distr.**: Western Ghats (Maharashtra). **Habit**: Common on forested hill slopes. Bright yellow flowers are borne among mauve-purple bracts. Flowers sometimes appear before the leaves. **Miscellaneous**: Food plant of Grass Demon Skipper butterfly. The small almond shaped tubers are boiled and eaten. **Flowering**: June-September.

220. INDIAN ARROWROOT *Hitchenia caulina*
Names: Hin, Beng: *Tikhur*, Mar: *Tavakhir, Chavar*. **Plant**: 1.2 m. **Flower**: 6 cm across; flower spike: 12-22 cm long. **Distr.**: Western Ghats. **Habit**: Endemic herb seen growing on hill slopes. Locally abundant. **Miscellaneous**: Food plant of Grass Demon Skipper butterfly. Tubers used to manufacture edible starch. **Flowering**: August-September.

MUSACEAE   **Banana Family**
221. WESTERN HILL BANANA *Ensete superbum*
Names: Mar: *Keli*. **Plant**: 3.5 m. **Flower**: 7.5 cm long; **Inflorescence**: drooping, up to 1.2 m long. **Distr.**: Western Ghats (Maharashtra to Kerala). **Habit**: Succulent, large herb, sprouts annually in the rains from its stout perennial rootstock. Seen among rocky hill slopes and cliffs, in open grassy places as well as forests. **Miscellaneous**: Rootstock and tender inflorescence are eaten. Young fruits are pickled. **Flowering**: July-November.

CANNACEAE   **Canna Family**
222. INDIAN SHOT *Canna indica*
Names: Hin: *Sabbajaya*, Mar: *Devkeli*, Beng: *Sarbajaya, Tel*: *Krishnatamara*, Tam: *Kalvalai*, Kan: *Hudingana*, Mal: *Kattuvala*. **Plant**: 1.2 m. **Flower**: 10 cm long. **Distr.**: Native of C. and S. America, naturalised in India, Sri Lanka. **Habit**: Gregarious, tall leafy herb, usually seen around ponds and marshes. Yellow and red-orange varieties are common. Black ballbearing shaped seeds used as ammunition, hence the common name. **Miscellaneous**: Sunbirds come for nectar and the pollinate flowers. **Flowering**: January-December.

AMARYLLIDACEAE   **Daffodil Family**

223. PINK STRIPED TRUMPET LILY *Crinum latifolium*
Names: Sans: *Madhuparnika, Vrishakarni*, Beng: *Sukhdarsan*, Tam: *Vishamungil.*
Plant: 90 cm. **Flower**: 6.5 cm across, 10 cm long. **Distr.**: India (absent in drier regions),
Sri Lanka. **Habit**: Large, showy, fragrant flowers appear during the first weeks of
monsoon. Prefers hilly and rocky terrain. Seen on the hills up to 800 m. **Miscellaneous**:
Food plant of Lily moth, whose brightly spotted caterpillars feed on the plant. Bulbs
are used in traditional medicine. **Flowering**: May-June.

224. FOREST SPIDER LILY *Pancratium parvum*
Plant: 12-20 cm. **Flower**: 5-10 cm long. **Distr.**: Western Ghats (Maharashtra,
Karnataka). **Habit**: This endemic is seen during the first few weeks of the monsoon.
Short-lived flowers often appear before the leaves. Prefers forested regions. **Flowering**:
June-July.

HYPOXIDACEAE   **Orchid-lily Family**

225. YELLOW GROUND STAR *Curculigo orchioides*
Names: Hin: *Kali musli*, Mar: *Kali musali*, Beng: *Talamuli*, Tel: *Nelatygadda*, Tam:
Nilappanaikkilanku, Kan: *Netatali-gadde*. **Plant**: 15-30 cm. **Flower**: 1.2 cm across.
Distr.: Peninsular India up to W. Bengal, Himalayan foothills from Kumaon eastwards
up to Northeast, Sri Lanka. **Habit**: Common on forest floors up to 1,600 m in the hills
at the onset of monsoon. Above 1,200 m, leaves more hairy and flowers deeper yellow.
Miscellaneous: Rhizome is used in traditional medicine. **Flowering**: July.

LILIACEAE   **Lily Family**

226. SPINY ASPARAGUS *Asparagus racemosus*
Names: Sans, Beng: *Satamuli*, Hin: *Satavar*, Mar, Guj: *Shatawari*, Raj: *Narkanto*, Tel:
Challagadda, Tam: *Shimai-shatavari*, Kan: *Majjige-gadde*, Mal: *Shatavari*. **Plant**: 6-
10 m. **Flower**: 6 mm across. **Distr.**: Throughout Subcontinent (absent in Bangladesh).
Habit: An armed vine with erect spines, seen in forested and drier regions and on the
hills up to 1,400 m. Flowers are fragrant. **Miscellaneous**: Flowers attract bees and
wasps. Seeds are dispersed by birds. Cultivated on a large scale for medicinal purposes,
and as an ornamental plant in gardens. **Flowering**: September-February.

227. EDIBLE CHLOROPHYTUM *Chlorophytum tuberosum*
Names: Hin: *Safed musli*, Mar: *Kuli*, Mal: *Veluttanilappana*. **Plant**: 20 cm. **Flower**:
2.5 cm across. **Distr.**: India, central and peninsular. **Habit**: One of the lilies that flower
with the first few showers, as the monsoon, breaks. Often seen in gregarious clumps, on
hills and forest edges. **Miscellaneous**: Tuberous roots are used in traditional medicine.
Leaves sold as monsoon produce are cooked and eaten. Generic name is derived from
Greek *chloros* (green) and *phyton* (a plant). **Flowering**: June-July.

228. GLORY LILY *Gloriosa superba*
Names: Sans: *Agnishikha*, Hin: *Kalihari*, Mar: *Kal-lavi*, Guj: *Dudhivachnag*, Beng:
Bishalanguli, Tel: *Adavinabhi*, Tam: *Kalaippaik-kilangu*, Kan: *Agnisikhe*, Mal: *Medoni*.
Plant: Up to 4 m. **Flower**: 8-10 cm across. **Distr.**: India (absent in drier regions), Sri
Lanka. **Habit**: Annual climber with leaf tips extending into tendrils. Seen in well-
wooded regions up to 500 m on the hills. Flowers change colour. **Miscellaneous**: Over-
collection of tubers for alkaloid extraction has endangered this plant. Tubers poisonous.
Used in traditional medicine. **Flowering**: August-September.

229. PALE GRASS-LILY *Iphigenia pallida*
Plant: 20-40 cm. **Flower**: 8 mm across. **Distr.**: Peninsular India (Western Ghats in Maharashtra-Karnataka, Deccan hills and Vidarbha region of Maharashtra). **Habit**: An annual herb with grass-like leaves, seen among open grassy meadows at the onset of the monsoon. One to four flowers, may be white or tinged with purple. This lily is endemic to the region. **Flowering**: June-August.

230. SOUTH INDIAN SQUILL *Scilla hyacinthina* (= *S. indica*)
Names: Hin, Beng: *Safed khus*. **Plant**: 12 cm. **Flower**: 6 mm across; flowering stalk: 20 cm long. **Distr.**: India, central and peninsular; Sri Lanka. **Habit**: Common herb, sprouting from underground tubers with the first rains. Leaves often have dark blotches. Fairly common in and around forest tracts, from plains to coast, up to 1,400 m on the hills. **Miscellaneous**: Flowers are pollinated by nectar-seeking bees and wasps. Tubers used in traditional medicine. **Flowering**: May-July.

PONTEDERIACEAE   **Water Hyacinth Family**

231. WATER HYACINTH *Eichhornia crassipes*
Names: Hin: *Jal kumni, Pishach kumbhi*, Mar: *Kendal*, Beng: *Kachuri pana*, Tel: *Pisachi thamara*, Tam: *Akasa thamarai*, Mal: *Kolavazha*. **Plant**: 15-25 cm. **Flower**: 5 cm across, flower spike: 15 cm long. **Distr.**: Native of tropical South America, naturalised in India. **Habit**: Brought to Asia for its pretty flowers, the plant is now a major menace, clogging waterways and larger wetlands. An aggressive coloniser that multiplies prolifically. All attempts to eradicate it have failed. **Flowering**: January-December.

232. LESSER WATER HYACINTH *Monochoria vaginalis*
Names: Hin: *Indivar*, Beng: *Nukha*, Tel: *Nirkancha*, Mal: *Kakapola*. **Plant**: 70 cm. **Flower**: 2 cm across. **Distr.**: India (including Himalaya), Nepal, Sri Lanka. **Habit**: An aquatic herb with spongy rootstock. Found in paddy fields, ditches, margins of ponds and marshes. Seen on the hills up to 1,500 m. **Miscellaneous**: Entire plant except roots is eaten as a vegetable. Leaves and roots are used in traditional medicine. **Flowering**: August-March.

COMMELINACEAE   **Spiderwort Family**
233. GARDEN COMMELINA *Commelina benghalensis*
Names: Hin: *Kanchara*, Mar: *Kena*, Raj: *Bakhana*, Tam: *Kanavazhar*. **Plant**: 60-90 cm. **Flower**: 10-15 mm across. **Distr.**: India, Nepal, Sri Lanka. **Habit**: One of the commonest among wayside monsoon annuals. Often seen in untended gardens. Flowers are sometimes white. Strangely, leafless underground suckers bear scapes of self pollinating white flowers which never open, but bear better seeds than those from the normal blue flowers. **Miscellaneous**: Leaves and rhizomes are cooked and eaten. Used to indicate presence of sulphur dioxide as air-pollutant. **Flowering**: August-December.

234. BEARDED COMMELINA *Commelina forskalaei*
Names: Hin: *Kana*. **Plant**: 10-15 cm. **Flower**: 8-11 mm across. **Distr.**: India (Maharashtra, Gujarat, Tamil Nadu, Delhi, Rajasthan), Pakistan. **Habit**: A slender, branching herb, more common in drier regions. Grows erect in gregarious, sheltered situations. Like Common Garden Commelina, it roots at nodes and produces several underground flowers on leafless basal branches. Spathes having long spreading hairs, give its common name. **Miscellaneous**: Cooked as vegetable. **Flowering**: August-November.

235. CRESTED CAT EARS *Cyanotis cristata*
Plant: 10-45 cm. **Flower**: 6 mm across. **Distr.**: India (including Himalaya), Sri Lanka.
Habit: A slender, erect herb, roots at basal nodes. Appears during rains among grasses
or in gregarious clumps from coast up to 1,000 m on the hills. Occurs sparsely on the
plains. **Flowering**: August-November.

236. FLUFFY CAT EARS *Cyanotis fasciculata*
Names: Mar: *Nirphulli*, Tel: *Golagandi*, Tam: *Nirupalli*, *Vallukkeippul*. **Plant**: 10-25 cm.
Flower: 5 mm across. **Distr.**: Peninsular India, Sri Lanka. **Habit**: This small, gregarious
herb is unmistakable, with its bright little flowers. The plant is less hairy in moist
situations and turns coppery as moisture reduces. Prefers moist, rocky, gravelly soil.
Grows from coast to plains and on the hills up to 750 m. **Flowering**: August-January.

237. GREATER CAT EARS *Cyanotis tuberosa*
Plant: 15-90 cm. **Flower**: 0.8-1 cm across. **Distr.**: Peninsular India, Sri Lanka. **Habit**:
This stocky, succulent herb is seen among the rocks in gravelly or sandy soil. Grows on
the plains in scrub forest, but prefers hills up to 800 m. **Miscellaneous**: Generic name
derived from Greek *kyanos* (blue) and *otus* (an ear) referring to the ear-like, blue petals.
Roots are used in traditional medicine. **Flowering**: August-September.

238. CREEPING CRADLE PLANT *Tonningia axillaris*
Names: Hin: *Kana*. **Plant**: 15-45 cm. **Flower**: 7 mm across. **Distr.**: India, Nepal, Sri
Lanka. **Habit**: Creeping succulent herb with erect tops seen in small clumps. Flowers
are cradled, as the name suggests, in the axillary position between the leaves and stem.
Locally abundant, it is seen from plains to hills up to 1,000 m. **Flowering**: August-
October.

ARACEAE Arum Family

239. DRAGON STALK YAM *Amorphophallus commutatus*
Names: Hin: *Jungli suran*, Mar: *Shevla*. **Plant**: 30-90 cm. **Flower**: 2.5-5.1 cm long;
flower stalk: 30-90 cm long. **Distr.**: Western Ghats (Maharashtra). **Habit**: Solitary
flower stalks appear, often in groups, just before the onset of the monsoon. Leaves
appear with the rains. Flies attracted by the fetid smell of flowers pollinate them.
Miscellaneous: Fed on by caterpillars of the Striped Green Hawkmoth (*Rhyncholaba
acteus*) and Silver Striped Hawkmoth (*Hippotion celerio*). Flower stalks are sold by
tribals as monsoon produce. Generic name is derived from Greek *amorphos* (deformed)
and *phallus* referring to the shape of the tubers. **Flowering**: May-June.

240. TARO *Colocasia esculenta*
Names: Sans, Beng: *Kachu*, Hin: *Arvi*, Mar: *Alu*, Tel: *Chemagadda*, Tam: *Seppan-
kizhangu*, Kan: *Shamagadde*, Mal: *Chembu*. **Plant**: 1.5 m. **Flower**: 20-35 cm tall.
Distr.: India, Sri Lanka. **Habit**: Gregarious herb growing from tubers on the banks of
streams, ponds, marshes and in moist and shady situations in forests. On the hills up to
1,400 m. Often cultivated. Plants may have green or purple leaf stalks. **Miscellaneous**:
Caterpillars of the Pinstriped Hawkmoth (*Theretra pinastrina*) feed on the plant.
Leaves, stems and tubers are cooked and eaten. **Flowering**: July-December.

GLOSSARY

Alternate – Leaf arrangement consisting of single leaves placed at different heights along a stem; the converse of opposite arrangement. This applies to all leaves whether they are simple or compound.

Annual – A plant that grows to maturity, flowers, and fruits in one year or one growing season and then dies.

Auricle – Ear-like flap at the base of a leaf.

Axil (Axillary) – The angle or point between the leaf and stem or branch.

Berry – A fleshy, rounded fruit with seeds.

Biennial – A plant that completes its flowering, fruits and dies in two years.

Binomial – A name consisting of two parts: genus or generic name, and species, or specific name.

Bract – A modified leaf below a flower or an inflorescence.

Bristle – A long stiff hair.

Bulb – A subterranean and modified leafy stem serving as a storage organ and consisting of overlapping leaf and often flower buds.

Bulbil – A small bulb produced on stems or in inflorescences and serving the same function as the larger subterranean bulbs.

Bulbous – Having a large bulb or in the shape of a bulb.

Calyx – The outer ring of the flower consisting of sepals.

Compound – A compound leaf has several or many distinct leaflets.

Corolla – A collective term for the petals of a flower. In many flowers, the petals are not separate but united into a corolla tube.

Deciduous – A term used to describe trees or shrubs, whose leaves fall after a certain season. The opposite of "evergreen".

Diurnal – A term used to describe flowers that open in the daytime.

Endemic – A species whose natural range is confined to a certain described area.

Epiphyte – A plant that grows upon another plant, using it for support without invading the tissues of the host plant.

Evergreen – Plants which retain their green leaves throughout the year.

116

Family – A natural grouping of plant genera with certain essential characteristics in common.

Female flower – Flower with a fertile ovary but without fertile stamens.

Flower-head – Densely packed with a group of flowers.

Genera – The plural of **genus**.

Genus – A term used in classification of plants and animals for closely related species which have common features and characteristics. Several genera make up a Family.

Herb – A plant lacking woody parts, dying down after each season.

Himalaya – The entire range of mountains and foothills from Kashmir in the west to Arunachal Pradesh in the east.

Inflorescence – A cluster of flowers originating from a single point on the stem, branch or trunk.

Introduced – A plant brought into a region where it is not native.

Latex – Milky sap.

Leaflet – One of the distinct divisions (blades) of a compound leaf.

Liana – Woody climber.

Male flower – Flower with fertile stamens but no fertile ovary.

Mid-vein or midrib – Central vein of the leaf.

Native – Naturally occurring in a region, not introduced.

Naturalised – A species introduced from another region, now fully established and reproducing naturally.

Nocturnal – A term used to describe flowers that open at night.

Node – The point on a stem from which a leaf or group of leaves emerges.

Parasite – An organism feeding and living on another organism.

Perennial – A plant that lives for at least three years; usually denotes a herbaceous plant or a herbaceous shrub. Several perennials die back above the ground of a season but remain as dormant bulbs or tubers until the next growing season.

Petal – A modified leaf of the corolla. A flower consists of a series of highly modified leaves arranged in whorls. The outermost (lowest) whorl is collectively called "sepal" or "calyx" and its parts are usually leaflike. The next whorl is referred to as "petals" or "corolla" and its parts are less leaflike and usually coloured other than green. The calyx and corolla also may have fused parts or, in some flowers, one or both of these whorls may be completely lacking.

Petiole – The primary stalk of a simple or compound leaf.

Peduncle – Stalk of a flower cluster, each flower being carried on a pedicel.

Pedicel – Stalk of a single flower.

Pod – A long cylindrical or flattened fruit, usually splitting into two sides as in typical pea or bean.

Pollen – Spores containing male sex-cells of seed-plants.

Prickle – Soft spine or thorn.

Prostrate – Growing close to the surface of the ground.

Raceme – A simple, unbranched, elongated inflorescence with stalked flowers.

Rhizome – An underground or surface stem growing horizontally and giving rise to roots, stems, and leaves at its nodes or growing tips.

Rootstock – A synonym for "rhizome"or underground stem.

Rosette – Usually referring to a cluster of leaves or petals growing in an overlapping, circular or radial formation.

Runner – Aerial stems spreading above the ground and rooting at nodes to form new plants.

Sahyadris – See Western Ghats.

Scape – A leafless flower stem that rises directly from the ground (the root).

Sepal – Usually green, modified leaf, together forming the outer whorl or calyx to protect the bud.

Shrub – Perennial plant with woody stems, and branching from the base.

Simple – Describes a leaf that has but a single blade and is not divided into separate leaflets. Not compound.

Solitary – Flowers or other parts that are single, not in clusters.

Spathe – A relatively large and often coloured bract, surrounding or found at the base of an inflorescence.

Species – A basic unit in the classification of plants and animals. Members of a species that have common characteristics and breed freely among themselves to produce viable offspring.

Stamens – Male reproductive organs which bear pollen.

Standard – A large, erect petal, especially that of flowers of peas and beans.

Succulent – Thick and juicy.

Tepal – A form that is intermediate between sepal and petal, and not readily distinguished from either.

Terrestrial – Growing in the ground or soil.

Toothed – Small triangular saw-like projections of the margins of leaves or other organs.

Tuber – A relatively short, swollen part of the stem (as in potato) or root (as in dahlia), that is usually but not always subterranean. It has growing points from which emerge the shoots of new plants and serves as a food storage organ.

Tubercle – A small rounded protuberance, usually on a stem.

Variety (Var.) – Differing in minor characters or a race not sufficiently distinct to be counted as a species.

Vein – A small rib in the leaf.

Western Ghats – In the context of this book, refers to the entire range of hills facing the western coast from north to south, including Nilgiris, Palni, Anaimalai and associated hills of Karnataka, Tamil Nadu and Kerala in the south.

REFERENCES

Almeida, M.R. (1996-99), **Flora of Maharashtra**. Vols.1-2, Blatter Herbarium, Mumbai.

Almeida, S. M. (1990), **Flora of Savantwadi**. Vol. 1-2, Dehra Dun.

Anon (1948-1992), **The Wealth of India - Raw Materials**. Vols 1-11, Council of Scientific and Industrial Research, New Delhi.

Anon (1986), Useful Plants of India. Council of Scientific and Industrial Research, New Delhi.

Bhandari, M.M. (1990), **Flora of the Indian Desert**, MPS Repros, Jodhpur.

Bharucha, F.R. (1983), **A Textbook of the Plant Geography of India**, Oxford University Press, Bombay.

Bennet, S.S.R. (1987), **Name changes in Flowering Plants in India and Adjacent regions**, Triseas Publishers, Dehra Dun.

Bole, P.V. & M.R Almeida. (1981-85), **Material for the flora of Mahabaleshwar**, *J. Bombay. nat. Hist. Soc.* 77(3):430-64,78(3):548-85; 81(2): 364-79.

Bor, N.L. & M.S. Raizada. (1954), **Some Beautiful Indian Climbers and Shrubs**, Bombay Natural History Society, Bombay. Repr. 1999.

Cook, T. (1958), **Flora of the Presidency of Bombay**, Vols. 1-3, Botanical Survey of India, Calcutta.

Coombes, Allen J. (1992), **The Hamlyn Guide to Plant Names**. Reed International Books Ltd., London.

Dassanayake, M.D., F.R. Fosberg, *et al.* (1980-94), **A Revised Handbook to the Flora of Ceylon**, Vols. 1-8, Amerind Publishing Co., New Delhi.

Hooker, J. D. (1872-97), **The Flora of British India**, Vols. 1-7, L. Reeve, London.

Lucas, G. & H. Synge. (1978), **The IUCN Plant Red Data Book**, The International Union for Conservation of Nature and Natural Resources, Morges, Switzerland.

Maheshwari, J.K. (1963), **Flora of Delhi**, Council of Scientific and Industrial Research, New Delhi.

Maheshwari, J.K. (1966), **Illustrations to the Flora of Delhi**, Council of Scientific and Industrial Research, New Delhi.

Matthew, K.M. (1981-88), **Flora of the Tamil Nadu Carnatic**, Vols. 1-4, Rapinat Herbarium, Tiruchirapalli.

Matthew, K.M. (1991), **An Excursion Flora of Central Tamil Nadu, India,** Oxford & IBH, New Delhi.

Mistry, Manek (1990), **Flora of Ratnagiri District.** Vols.1-2, Ph.D. Thesis submitted to the University of Bombay.

Nayar, M.P. & A.R.K. Sastry. (1987-90), **Red Data Book of Indian Plants,** Vols.1-3, Botanical Survey of India, Calcutta.

Polunin, Oleg & A. Stainton. (1990), **Flowers of the Himalaya,** Oxford University Press, Delhi.

Rajbhandari, K.R. & R. Joshi. (1998), **Crop Weeds of Nepal,** Natural History Society of Nepal, Kathmandu.

Raut, Madhav (1959), **Our Monsoon Plants, Glimpses of Nature Series, No. 4,** Bombay Natural History Society, Mumbai.

Sahni, K.C. (1998), **Book of Indian Trees,** Bombay Natural History Society, Mumbai.

Saldanha, C.J. & D.H. Nicolson. (1976), **Flora of Hassan District Karnataka,** Amerind Publishing Co., New Delhi.

Santapau, H. (1967), **Flora of Khandala,** Records of the Botanical Survey of India, Vol. XVI, No.1. 3rd reprint, Govt. of India.

Shubhalaxmi, V. (1997), **Ecology of moths of Sanjay Gandhi National Park,** M.Sc. Thesis submitted to the University of Bombay.

Sebastian, Leela (1996), **Hedgerows or Highways,** Sanctuary, Vol. XVI, No.2: 47-48.

Ugemuge, N.R. (1986), **Flora of Nagpur District,** Nagpur.

Walter, K. S. & H. J. Gillett, eds. (1998), **1997 IUCN Red List of Threatened Plants,** International Union for Conservation of Nature and Natural Resources, Morges, Switzerland.

Index of Common Names

Index of Regional Names

Kanchara 112
Kandankathiri 84
Kandyuli 84
Kanghi 40
Kankero 60
Kankra 64
Kanputi 38
Kanta chaulai 102
Kanta jati 90
Kantakari 84
Kantalan-indranan 60
Kantalo bhende 40
Kantanotya 102
Kantanu-dant 102
Kantashelio 90
Kante math 102
Kante-indrayan 60
Kanthar 38
Kanthari 38
Kanti 44
Kantola 60
Kantoli 60
Kanval 36
Kaola 52
Kara 62
Karambal 94
Karambha 38
Karamcha 72
Karanjvel 50
Karanta 74
Kararmarda 72
Karaunda 72
Kare 62
Kareel 36
Karekayi 72
Kariecheera 38
Karikeerai 38
Karimpolan 58
Karira 36
Kariramu 36
Karithumba 100
Karitumbe 100
Kariyartharani 98
Kariyatu 88
Karlikai 60

Karpokkari 50
Karukanni 98
Karunta 90
Karuthellu 88
Karvanda 72
Karvi 92
Kasaunda 56
Kasittumbai 46
Kastula 92
Katte tumbe soppu 76
Katthurparathi 40
Kattu-mullen-keera 102
Kattu-thippali 98
Kattuvala 108
Kaundal 60
Kavach 52
Kavalai 96
Kazhuthaithumbai 76
Kebu 108
Keli 108
Kelia 50
Kena 112
Kendal 112
Kepala 64
Ker 36
Ker-gawl 80
Kerro 36
Kesaraja 66
Kesuti 66
Keu 108
Khadsampal 48
Khaj-kuili 52
Khajkhujli 52
Khar-indrayan 60
Khasa 88
Khat-kachario 60
Khatara 76
Khatari 44
Khattimeethi 46
Khetraubat-atibala 42
Khunkhunia 50
Khursa 38
Kidamari 102
Kidemar 102
Kilukilluppai 50

W

Y

Index of Scientific Names

NOTES

Munnar

Hypericum mysurense

Pittosporum neelgherrense

Rhodomyrtus tomentosa
(Myrtaceae)

Caesaria ovata (Flacourtiaceae)

Kalanchoe pinnata
(Crossulaceae)

Maesa indica (Myrsinaceae)

Leucas suffruticosa (Lamiaceae)